6th PING International Conference - Modern Trends in Material Engineering

IOP Conference Series: Materials Science and Engineering
Volume 1243

Plzen, Czechia
7 - 10 September 2021

Editors:

Ludmila Kucerova
Hana Jirkova
Stepan Jenicek

ISBN: 978-1-7138-5442-5
ISSN: 1757-8981

Printed from e-media with permission by:

Curran Associates, Inc.
57 Morehouse Lane
Red Hook, NY 12571

Some format issues inherent in the e-media version may also appear in this print version.

Institute of Physics
Dirac House, Temple Back
Bristol BS1 6BE UK

Phone: 44 1 17 929 7481
Fax: 44 1 17 920 0979

techtracking@iop.org

Additional copies of this publication are available from:

Curran Associates, Inc.
57 Morehouse Lane
Red Hook, NY 12571 USA
Phone: 845-758-0400
Fax: 845-758-2633
Email: curran@proceedings.com
Web: www.proceedings.com

TABLE OF CONTENTS

IOP Conf. Series: Materials Science and Engineering 1243 (2022) 011001 doi:10.1088/1757-899X/1243/1/011001

Preface

We would like to introduce the selected articles from the 6th PING Junior conference - Modern Trends in Materials Engineering, which took place from 7 to 10 September 2021 in Pilsen, Czech Republic. The conference is regularly organized by the Regional Institute of Technology (RTI, www.rti.zcu.cz) at the University of West Bohemia.

This year's event was focused mainly on master's and doctoral students to support their research activities and enable them to present the obtained results. Furthermore, selected research projects were presented within the contributions addressed within RTI. Emphasis was placed on the results of the project "Research of additive technologies for future applications in the machinery industry - RTI plus" (No. CZ.02.1.01/0.0/0.0/18_069/0010040).

This year was very demanding for organizing mass events, and our conference was also affected by the coronavirus epidemic. For this reason, it was organized in a combination of presentations with personal attendance and online participation. Online participation was through Google Meet.

A total of 19 contributions were presented at the conference. The 11 best papers were selected for publication in the IOP.

In conclusion, we would like to invite you to the seventh year of the PING conference, which will take place from 13 to 16 September 2021.

Organizing committee of PING 2021

Scientific Committee

Doc. Ing. Tomáš Křenek Ph.D., University of West Bohemia, Czech Republic
Doc. Ing. Ludmila Kučerová, Ph.D., University of West Bohemia, Czech Republic
Doc. RNDr. Dagmar Jandová Ph.D., University of West Bohemia, Czech Republic
Doc. Ing. Ladislav Socha, Ph.D., Institute of Technology and Business in České Budějovice, Czech Republic
Ing. Josef Odehnal, Ph.D., University of West Bohemia, Czech Republic
Dr. Ing. Hana Jirková, Ph.D., University of West Bohemia, Czech Republic
Ing. Štěpán Jeníček, University of West Bohemia, Czech Republic

Organizing Committee

Ing. Štěpán Jeníček, RTI, University of West Bohemia, Czech Republic
Doc. Ing. Ludmila Kučerová, Ph.D., RTI, University of West Bohemia, Czech Republic
Ing. Kateřina Opatová, RTI, University of West Bohemia, Czech Republic
Dr. Ing. Hana Jirková, Ph.D., University of West Bohemia, Czech Republic
Mgr. Dagmar Bublíková, RTI, University of West Bohemia, Czech Republic

Program Committee

Ing. Štěpán Jeníček, RTI, University of West Bohemia, Czech Republic
Doc. Ing. Ludmila Kučerová, Ph.D., RTI, University of West Bohemia, Czech Republic

Contacts

University of West Bohemia
Faculty of Mechanical Engineering – Regional Technological Institute – PING conference
Univerzitni 8
306 14 Plzen
Czech Republic
Phone: +420 377 638 762
E-mail: jeniceks@rti.zcu.cz

www.ping.zcu.cz

IOP Conf. Series: Materials Science and Engineering 1243 (2022) 011001 doi:10.1088/1757-899X/1243/1/011001

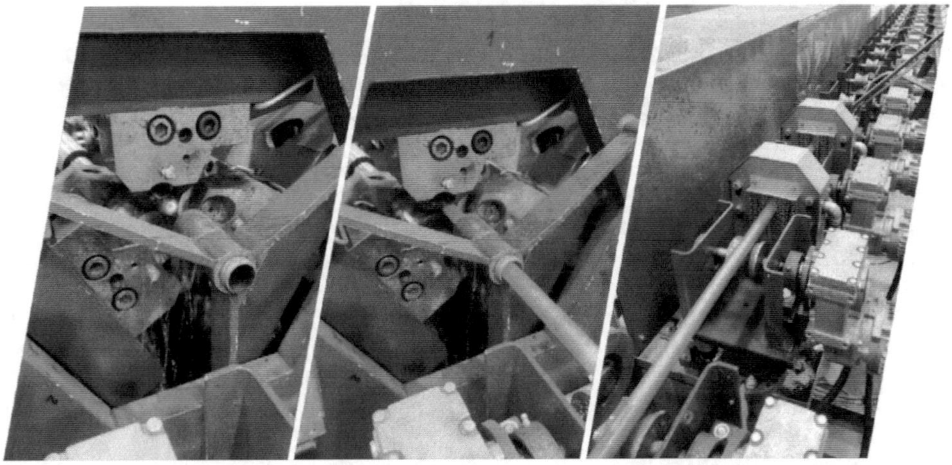

Proceedings PING Junior 2021 - Modern trends in material engineering

Editors of Proceeding
Doc. Ing. Ludmila Kučerová, Ph.D.
Dr.-Ing. Hana Jirková, Ph.D.
Ing. Štěpán Jeníček

PING Junior 2021 is organized with the support of funds for specific university research project SVK1-2021-008 and ERDF "Research of additive technologies for future applications in the machinery industry - RTI plus" (No. CZ.02.1.01/0.0/0.0/18_069/0010040).

EUROPEAN UNION
European Structural and Investment Funds
Operational Programme Research, Development and Education

MINISTRY OF EDUCATION, YOUTH AND SPORTS

This conference is organized under the patronage of the Regional Institute of Technology of the University of West Bohemia in Pilsen.

IOP Conf. Series: Materials Science and Engineering 1243 (2022) 011002 doi:10.1088/1757-899X/1243/1/011002

Peer Review Statement

All papers published in this volume have been reviewed through processes administered by the Editors. Reviews were conducted by expert referees to the professional and scientific standards expected of a proceedings journal published by IOP Publishing.

- **Type of peer review:** Single Anonymous
- **Conference submission management system:** Morressier
- **Number of submissions received:** 19
- **Number of submissions sent for review:** 17
- **Number of submissions accepted:** 10
- **Acceptance Rate (Submissions Accepted / Submissions Received × 100):** 52.6
- **Average number of reviews per paper:** 2.47
- **Total number of reviewers involved:** 19
- **Contact person for queries:**
  Name: Štěpán Jeníček
  Email: jeniceks@rti.zcu.cz
  Affiliation: University of West Bohemia

IOP Conf. Series: Materials Science and Engineering 1243 (2022) 012001 doi:10.1088/1757-899X/1243/1/012001

Analysis of the causes of undesirable curvature of the workpiece during longitudinal rolling using numerical simulation

S Benešová[1*], Karolína Burdová[1], J Hajšman[1]

[1]University of West Bohemia in Pilsen, Faculty of Mechanical Engineering

[*]E-mail: sbenesov@kmm.zcu.cz

Abstract Undesirable additional deformation was observed during the experimental cold rolling of steel. A simulation was performed in Deform software to clarify the causes. The influence of the unequal coefficient of friction, geometric inaccuracies of the rolls and the conveyor, and the situation when the rolls did not have the same diameter or rotated at different speeds were investigated. Simulations with different roll speeds corresponded best to reality. The simulation of the undesired curvature of the workpiece at different speeds of rotation of the rolls and 30% reduction was subsequently performed as an elastoplastic problem in order to determine the amount of springback and residual stress. Furthermore, the press straightening of the workpiece was simulated. Inhomogeneous and accidental deformation and residual stresses must be taken into account, especially when researching the effect of cold rolling on the mechanical properties and structure of the material.

1 Introduction

In cold forming, the resulting structural state and mechanical properties are a function of the input parameters of the material and the method of forming, where the main forming factors are the amount of deformation, stress state and a less known but important factor in multi-step forming, which is the "history" of deformation [1]. Compared to hot forming, the temperature has no effect and also the effect of strain rate is suppressed, since temperature and time-limited processes such as recrystallization and recovery do not take place in the material. The deformation is inhomogeneously distributed in the workpiece after forming by conventional procedures (rolling, forging). A side effect of inhomogeneous deformation inside the workpiece is residual stress. The stress state for a given material is determined by the shape of the deformation zone and the amount of friction. The factor of the history of deformation is related to the physical nature of the dislocations movement, their accumulation and the formation of crack nuclei. Compressive stress closes microcracks resulting in improved material formability, on the contrary, the tensile stress opens the crack and soon leads to plasticity depletion and fracture [2]. An example can be found in the technology of gradual drawing of wire in dies. For the material with the initial elongation of about 10% the following coefficient of elongation is achieved:

$$\lambda = \frac{l_1}{l_0} \approx 15 \quad (1)$$

where l_1 is the final length after several operations and l_0 is the initial length of the blank [3]. During experimental cold rolling on a quarto-type rolling mill [4], undesired significant rounding of the workpiece occurred. Thus, an additional bending deformation was introduced into the material, where one side is subjected to tensile stress and the other compressive. Subsequently, this workpiece was mechanically press-straightened between the flat plates. During this operation, additional deformation is introduced into the workpiece and with it unwanted residual stress. If the aim is to study the structural processes that are related to the cold deformation of the material, then this additional deformation must be avoided or at least its size and distribution must be defined.

IOP Conf. Series: Materials Science and Engineering 1243 (2022) 012001 doi:10.1088/1757-899X/1243/1/012001

2 Numerical simulation of longitudinal rolling

In the first phase, the simulation was performed under precise conditions with a 30% reduction. Rolling parameters which correspond to experimental practice are summarized in tab. 1.

Tab 1. Rolling parameters

Roll diameter	240 mm
Roll surface velocity	0,2 m/s
Billet size (length, width, thickness)	335 x 125 x 21,5 mm
Reduction	30 %

The deformation characteristics of the material were taken from the Deform database for chrome-nickel stainless steel DIN X10CrNiMoTi 18-10 with an initial yield strength of approx. 360 MPa and a maximum yield strength after strengthening of 1050 MPa. The coefficient of friction was defined by Coulomb's law and was set to 0.8. The coefficient of friction is relatively high compared to different forming operations. For rolling, a high coefficient of friction is necessary for the billet to enter the gap between the rolls. For rolling on smooth rolls following conditions must be met:

$$f > tg\,\alpha \qquad (2)$$

where f is the coefficient of friction and α is the angle of bite [5]. The simulation of precision rolling is shown in Fig. 1. In accordance with the rolling theory, the deformation is delayed in the central region of the workpiece, which causes the periodicity of the deformation distribution in the workpiece.

3 The causes of undesirable curvature of the workpiece

Several options of incorrect rolling mill settings have been identified. The effect of the workpiece was not considered, since the same manner of upwards curving around the upper roll is always observed when the workpiece is inserted between the rolls.

3.1 Difference in friction between the workpiece - upper roll and workpiece - lower roll interfaces

The coefficients of friction were set as follows: workpiece - upper roll f = 0.9, workpiece - lower roll f = 0.7. The situation is shown in Fig. 2. The curvature is small and the additional undesirable deformation cannot be expected to affect the structure. Additional deformation accumulates in surface areas (red colour).

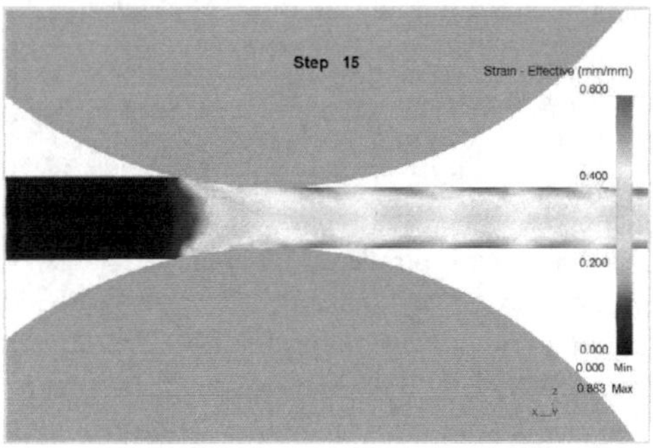

Fig. 1 Effective strain at 30% reduction during precision rolling

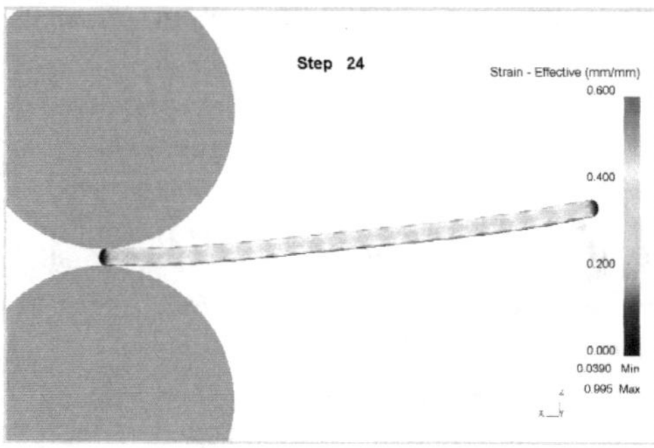

Fig. 2 Shape of the workpiece for different friction coefficients between the workpiece and the rolls

3.2 Inaccurately adjusted conveyor, billet moved upwards by 2 mm in the longitudinal axis

The situation is shown in Fig. 3. There is a significant downward curvature, which does not correspond to reality. When curved downwards, the rolled blank would be trapped on the conveyor and the roll could be jammed or forcibly straightened on the conveyor.

3.3 Inaccurately adjusted conveyor, billet shifted downwards in the longitudinal axis by 2 mm

The billet is bent when pulled between the rolls, it is pushed against the conveyor plate resulting in an undesirable local deformation and cutting off the surface of the billet in one place. Later, the rolled blank is lifted by the roll and the result is a relatively straight workpiece.

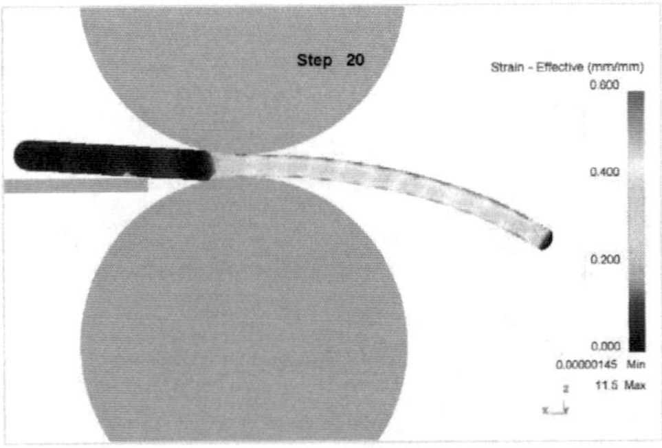

Fig. 3 Conveyor moved 2 mm upwards - above the rolling axis

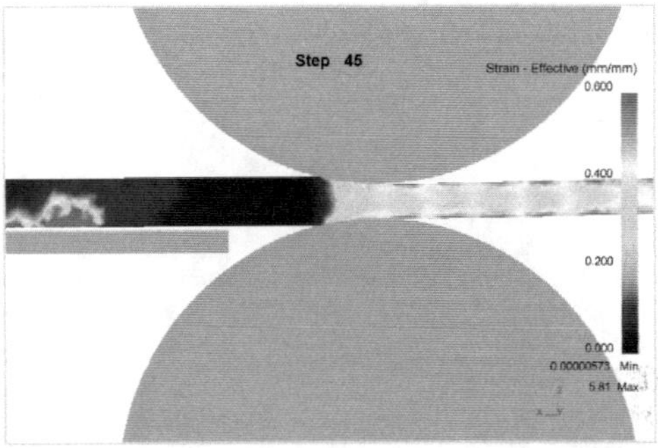

Fig. 4 Conveyor moved 2 mm downwards – beneath the rolling axis

3.4 Inaccurate roll mounting - the upper roll is advanced in front of the lower roll

If the centres of the rolls are not on the same vertical axis, the billet first comes into contact with one roll. If the upper roll is moved in front of the lower roll, contact with the upper roll will occur first. Fig. 5 shows a situation where the centre of the upper cylinder is advanced by 5 mm. This will cause a slight downward curvature of the billet.

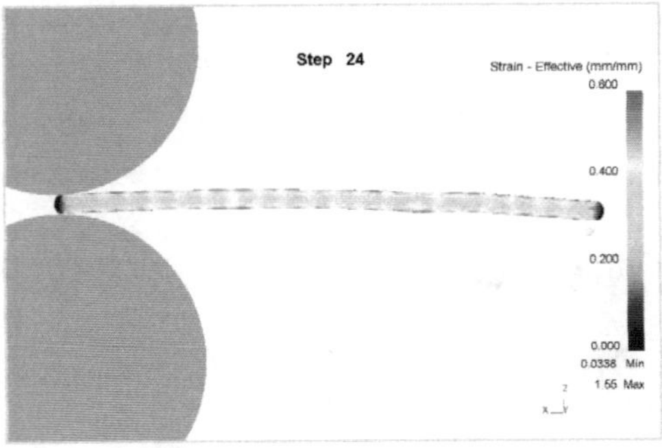

Fig. 5 Inaccurate roll mounting – the upper roll advanced by 5 mm forward

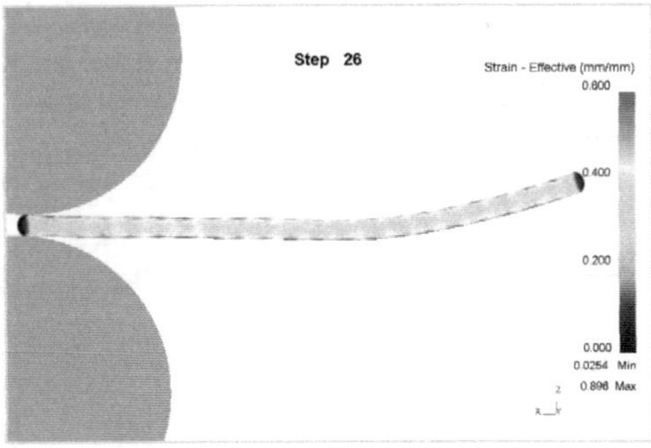

Fig. 6 Inaccurate roll mounting – upper cylinder moved backwards by 10 mm

3.5 Inaccurate roll placement - upper roll moved backwards
If the centre of the upper roll is moved backwards, the billet first comes into contact with the lower roll. Fig. 6 shows the situation after rolling when moving the centre of the upper roll by 10 mm. The billet is curved at the beginning of rolling after being pulled between the rolls. Later the rolling stabilize and further additional deformation is negligible.

3.6 *Unequal roll diameters*
This defect of the rolling mill is unlikely and easily measurable. In the simulation, the diameter of the lower roll was 240 mm, the upper 236 mm. Fig. 7 shows a significant curvature around the upper roll, which is similar to what was actually observed.

3.7 Unequal roll velocity
In the simulation, the angular velocity of the upper cylinder was set to 1.66 rad / s, which corresponds to 0.265 rps and the peripheral velocity of the cylinders 2 m/s. The lower cylinder had an angular velocity of 1.7 rad / s and 0.27 rps. The result is shown in Fig. 8. The deformation is inhomogeneous, the curvature is large, and there is a combination of reduction and bending.

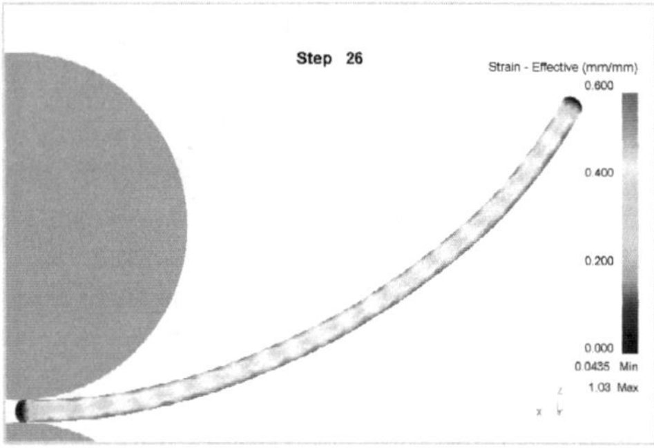

Fig. 7 The upper roll diameter is smaller than the lower

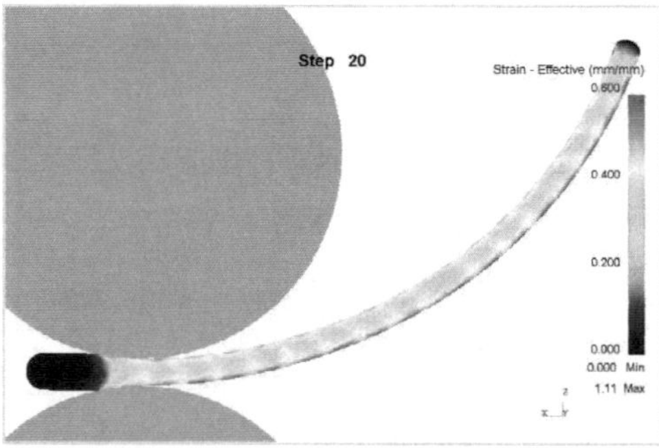

Fig. 8 Unequal rotation speed - lower roll speed is higher

4 Workpiece straightening by pressing between flat plates
Significantly curved billets cannot be further rolled. Therefore, they were press-straightened between the plates, thus further cold deformation was introduced into the billet. Fig. 9 shows the outer side of the rolled piece for the case of different roller rotation speeds (chapter 3.7) after straightening under a press with a high deformation over the entire surface.

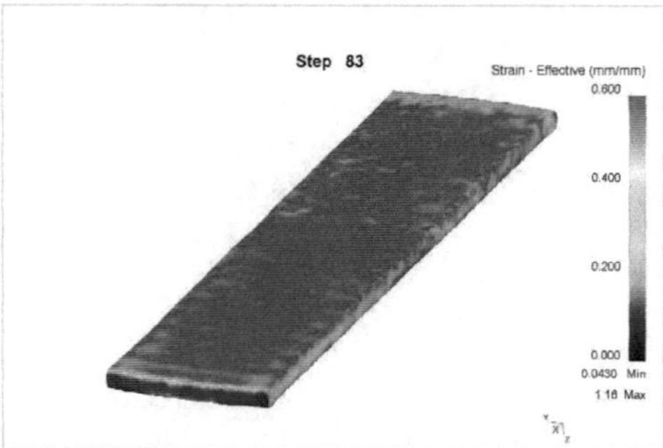

Fig. 9 Condition after straightening between the plates - the outer side of the originally bent workpiece

5 The springback simulation

Significant springback occurs during cold forming. Therefore, a simulation was performed in which the rolled body was defined as an elasto-plastic object. When performing a simulation with unequal rotation speed according to chap. 3.7, when the upper roll rotated at 1.66 rad/s (0.265 rpm), the lower cylinder 1.7 rad/s (0.27 rpm) with the springback, the shape of the roll was significantly less curved compared to the experiment. Based on a consultation with the technologist, it was suggested that due to the construction specifics of the rolling mill, the upper roll is delayed during engagement with respect to the set speed. Therefore, in the simulation the upper roll speed was set to 1.5 rad / s (0.24 rps), the lower roll speed was set to the standard rolling mill speed of 1.66 rad / s (0.265 rps). Subsequently, the press-straightening and the unloading were simulated. The residual stress on the outer side is inhomogeneously distributed and reaches up to 1000 MPa, i.e. values just below the yield point. The shape of the straightened workpiece, the level of internal stress and its distribution after unloading are shown in Fig. 10 and Fig. 11.

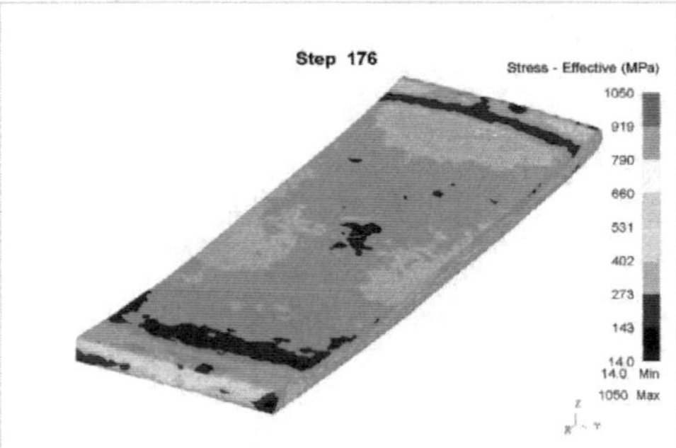

Fig.10 The shape after straightening – residual stress after the springback. Very inhomogeneous distribution.

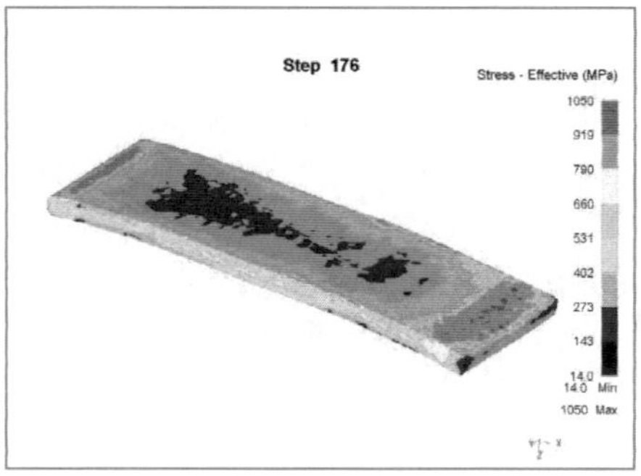

Fig. 11 The shape after straightening – residual stress after the springback. The bottom side of the workpiece.

6 Conclusion

In the experiment, during the cold rolling of a 335 x 135 x 21.5 mm austenitic steel billet, there was a significant curvature around the upper roll. Possible causes were analyzed by numerical simulation, namely:

- inaccurate rolling mill geometry and inaccurate feeding of the billet between the rolls
- unequal friction
- different roll diameters
- different rotation speeds of the rolls

It was shown that:

- Inaccurate rolling mill geometry and inaccurate feeding of the billet between rolls will cause inhomogeneous deformation along the workpiece, which is different from the real observation
- The different friction, when the coefficient of friction between the lower roller and the billet was 0.7 and between the upper roller and the billet, caused the curvature of the workpiece similar to that observed in the experiment, but to a lesser extent. Friction depends on the quality of the roll and billet surfaces. If the top and bottom of the billet have different surface roughnesses, this situation is likely. In fact, there is always an upward curvature - a rounding around the upper cylinder. If the blank is randomly inserted between the rolls, the curvature would have to occur randomly upwards or even downwards, which would result in blocking of the conveyor at the exit of the rolls. It is advisable to check the surface quality of the rolls.

- Different cylinder diameters (240 and 236 mm) lead to significant curvature, nevertheless, this situation is unlikely. It is recommended to remeasure the cylinder diameters.
- The situation when the upper cylinder rotates more slowly than the lower one appears to be the closest to the real observation. During rolling, there is a combination of rolling with reduction and bending. Greater deformation occurs on the outer - longer side of the bent workpiece. When setting the rolled plate as an elastoplastic body, springback was taken into account.

By press-straightening, an additional deformation is introduced, which accumulates on the outer side of the originally bent workpiece. This results in uneven strengthening of the surface and subsurface layers. The residual stress is unevenly distributed, with some areas almost unloaded, while elsewhere the residual stress reaches almost the yield point, in our case, it was up to 1000 MPa. This cannot be neglected, especially in the research of the influence of the amount of deformation on structural processes, e.g. during special heat treatment. Inhomogeneous deformation results in the occurrence of residual stress, which affects physical-metallurgical processes, such as the progress of phase transformations. The cause of bending and rounding should be found in the adjustment of the rolling mill. In particular, in the case of a quarto rolling mill, where the support rollers are driven, the upper working cylinder may slip and be delayed. It is also possible that the upper cylinder slows down during engagement due to gear wear, bearing play and other design inaccuracies. The conclusions of these simulations were provided to the technologists.

Acknowledgements
This research was funded by the Technology Agency of the Czech Republic under the project TJ04000265 „Optimization of properties of maraging steel with reverse austenite and TRIP effect".

References

[1] Wagoner R H and Chenot J 1997 Fundamentals of metal forming. John Willey&Sons
[2] Petruželka J, Sonnek P 2007 Tvařitelnost kovových materiálů. VŠB Ostrava
[3] Benešová S 2007 Materiálové modelování a numerická simulace jako nástroj pro vývoj technologických procesů. Dizertační práce, Západočeská univerzita v Plzni
[4] Semiatin S L 2005 Metalworking: Bulk forming. Vol. 14A. ASM International.
[5] Žídek M, Kuře F 1986 Válcování. 2. vyd. VŠB Ostrava

Ballistic Resistance of ARMOX Advance according to NATO standard AEP-55 STANAG 4569

V Coufal[1*], J Křesťan[2] and M Vnouček[3]

[1]Západočeská univerzita v Plzni, Univerzitní 8, 30614 Plzeň,

[2]VVÚ, s. p., Veslařská 230, 63700 Brno

[2]ŠKODA JS a.s., Orlík 266/15, Pilsen, Czech Republic, EU

[*]E-mail: coufalvlada@tiscali.cz

Abstract. ARMOX low-alloy high-strength steel sheet is used, among other things, as a basic homogeneous ballistic protection in plate carriers and vehicle bodies. The hardest armour plates of the ARMOX series include ARMOX Advance, which, even with a smaller thickness, provides the same ballistic protection as the stronger ARMOX 500 or ARMOX 600 sheets. The manufacturer of ARMOX armour plates declares for the ARMOX Advance with a thickness of 5 mm the resistance to small arms fire with 7.62x51 and 5.56x45 rounds according to NATO AEP-55 STANAG 4569 Annex A level 1 and for the thickness of 7 mm the ballistic resistance to small arms fire with 7.62x39 API BZ rounds according to NATO AEP-55 STANAG 4569 Annex A level 2. Due to the manufactured thicknesses, ballistic resistance is not specified for levels 3 to 6 in accordance with NATO AEP-55 STANAG 4569 Annex A. For levels 1 to 3, STANAG 4569 does not require a test against flying fragments from 155 mm artillery shells, but it is an optional component at the discretion of the national authority. In order to be able to use this type of sheet for this purpose, empirical equations were used to determine whether the conditions set out in STANAG 4569 Annex A Level 1 and Level 2 are met over the full range of ballistic resistance required. From the results obtained it was found that the ballistic resistance of ARMOX Advance armour plate against flying artillery shell fragments is not met and therefore it cannot be used alone as a basic homogeneous ballistic protection in vehicle bodies and protective sheet carriers to provide protection of persons in areas where artillery fire is expected.

1 Introduction

The Swedish company SSAB is one of the world's leading manufacturers of ballistic resistant steels, which are currently used to produce ballistic resistant materials for plate carriers and bodies of civilian and military vehicles. The use of alloying elements and modern methods of forming materials reduces the thickness of the plates while maintaining ballistic protection for people and vehicle crews, thus increasing the mobility and range of the equipment. The level of ballistic protection of personnel and lightly armoured military vehicles against projectile penetration is specified in NATO AEP-55 STANAG 4569 Annex A, which currently consists of six levels, each defining the types of weapons and munitions against which the vehicle body must protect the crew (Table 1). In addition, and also for reference purposes, resistance to 155 mm artillery shells is also defined (the levels differ essentially in the distance of explosion and the angle of impact of the shell on the armour) [1].

Manufacturers and sellers declare for the NATO standard AEP-55 STANAG 4569 Annex A level 1 and level 2 ballistic resistance of armoured plates only when using firearms (Table 2), i.e. they do not guarantee the protection of the crew of armoured vehicles in places where shelling by artillery shells of 155 mm calibre can be expected [2].

Table 1. STANAG 4569 Ed. 2 – Degree of resistance of armoured vehicle [1].

Article I. Level	Article II. Weapons, ammunition and conditions	Article III. Fragments from 155 mm grenades, simulation with ammunition
1	Assault rifles 7.62mm×51 NATO ball (833 m/s) 5.56mm×45 NATO SS109 (900 m/s) 5.56mm×45 NATO M193 (937 m/s) Distance 30 m, angle of incidence 0° to 30°	Distance 100 m, angle of incidence 0° to 18° 12.7mm: 310 m/s 20mm: 520 m/s
2	Assault rifles 7.62mm×39 API BZ (695 m/s) Distance 30 m, angle of incidence 0° to 30°	Distance 80 m, angle of incidence 0° to 22° 12.7mm: 420 m/s 20mm: 630 m/s
3	Assault and sniper rifles 7.62mm×51 AP WC (930 m/s) 7.62mm×54R B32 API (854 m/s) Distance 30 m, angle of incidence 0° to 30°	Distance 60 m, angle of incidence 0° to 30° 12.7mm: 560 m/s 20mm: 770 m/s
4	Heavy machine guns 14.5mm×114 API/B32 (911 m/s) Distance 200 m, angle of incidence 0°	Distance 25 m, angle of incidence 0° to 90° 20mm: 960 m/s
5	Automatic cannons 25mm×137 APDS-T PMB 073 (1258 m/s) Distance 500 m, angle of incidence 0°	Distance 25 m, angle of incidence 0° to 90° 20mm: 960 m/s
6	Automatic cannons 30mm x 165 AP-T (810 m/s) 30mm x 173 APFSDS-T (1330 m/s) Distance 500 m, angle of incidence 0°	Distance 10 m, angle of incidence 0° to 90° 20mm: 1250 m/s

Note: Shot impact velocities are given. Tolerance ±20 m/s.

Table 2. Declared resistance of ARMOX Advance armour plates [2].

		Article IV. STANAG 4569				
Level	Ammunition	Speed (m/s)	Distance (m)	Armox 500 (mm)	Armox 600 (mm)	Armox Advance (mm)
1	7.62mm×51 5.56mm×45 SS109 5.56mm×45 M193	833 ±20 900 ±20 937 ±20	30	9	6	5
2	7.62mm×39 API BZ	695 ±20	30	12	8	7

The work was focused on the description and determination of the required thickness of the ARMOX Advance armour plate, which will lead to the provision of ballistic protection of armoured vehicles against 5.56x45,7.62x51 and 7.62x39 API BZ projectiles and fragments from 155 mm calibre artillery shells, thus ensuring the protection of vehicle crews according to NATO AEP-55 STANAG 4569 Annex A level 1 and level 2 standards in the entire range of ballistic resistance (mandatory and optional components of the standard).

2 Experimental program

ARMOX Advance armour plates were used for the experimental program. Plates are manufactured in thicknesses from 4.0 to 7.9 mm with a tolerance of -0.0 to +0.6 mm and a declared hardness of 58 to 63 HRC. The chemical composition of the ARMOX Advance armour plate declared by the manufacturer is given in Table 3 [3]. The microstructure of the as-delivered condition consisted of a low tempered fine-grained martensite (Figure 1).

Table 3. Chemical composition of ARMOX Advance armour plate [wt. %] [4].

C	Si	Mn	P	S	Cr	Ni	Mo	B
0.74	0.7	1	0.01	0.003	1.5	3	0.7	0.005

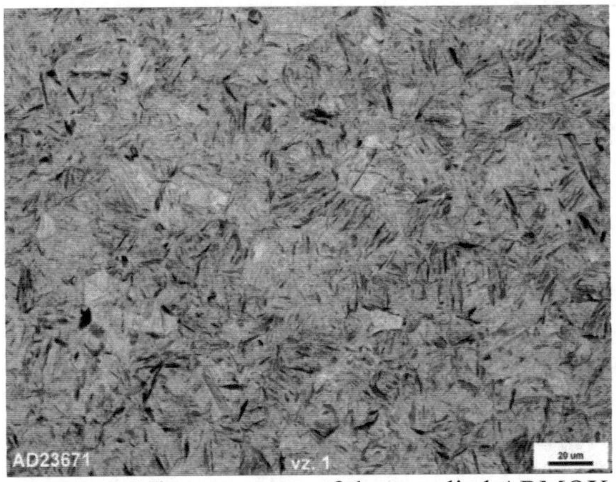

Figure 1. Microstrucutre of the supplied ARMOX Advance armour plate.

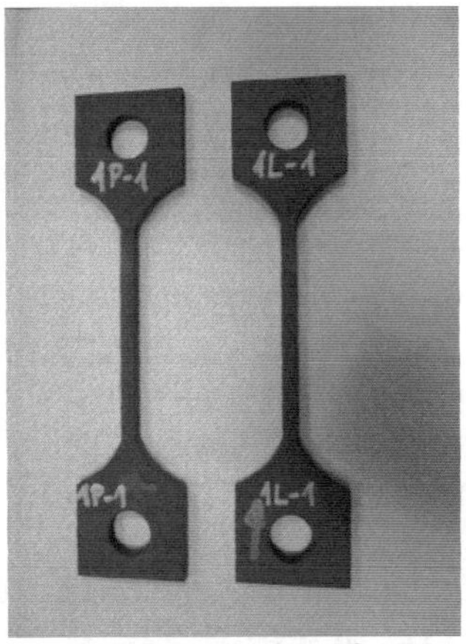

Figure 2. Test samples of ARMOX Advance steel.

The manufacturer does not provide any information on the mechanical properties of ARMOX Advance steel. In order to obtain basic information on mechanical properties, ARMOX Advance 5 mm thick plate was used and subjected to tensile testing according to ČSN EN ISO 6892-1: Metallic materials - Tensile testing and hardness testing.

The anisotropy of the material was determined on specimens cut in the longitudinal (L) and transverse (P) directions relative to the rolling direction (Figure 2). The test results are shown in Table 4. Better results of the tensile test were found for specimens whose fibres were oriented in the longitudinal direction to the loading direction. The rolling direction has an effect on the mechanical properties of the material and on different types of forming, e.g. bending.

Table 4. Tensile test results according to ČSN EN ISO 6892-1.

Sample	$R_{p0,2}$ [MPa]	R_m [MPa]	A [%]	Z [%]
1P	1465.5	2372.5	5.205	13.47
1L	1492	2458.67	6.71	19.51

The Rockwell and Brinell methods were chosen for the hardness test. The Rockwell hardness test was performed to compare the actual hardness of the material with the hardness declared by the manufacturer. The measured hardness (Table 5), although at the lower end of the declared hardness range, meets the hardness of the ARMOX Advance armour plate as declared by the manufacturer. The Brinell hardness test with a sintered carbide ball (Table 5) was performed to verify the use of the Gebeaud equation to determine the limiting impact velocity of armour perforation by projectiles of a different calibre.

Table 5. Average hardness from the hardness test.

Hardness test	HRC	HBW
Average hardness	58,7 ± 0.3	633,8 ± 2.0

In order to determine suitable plate thicknesses, it was necessary to calculate the impact energies of the projectiles and fragments. The impact energy of the projectile is calculated using the formula:

$$E_d = \frac{1}{2}.m.v^2 \text{ [J], where} \tag{1}$$

m (kg) is the mass of the projectile and v (m.s^{-1}) is the velocity of the projectile required to perforate the armour plate, which can also be defined as the limiting impact velocity [4].
In order to find the thickness of the plate at which perforation does not occur, a formula was used where the impact energy is equal to the work done to create a hole of the projectile diameter in the plate of a given thickness $E_d = W$, where the work is calculated according to the formula:

$$W = F.s \quad \text{[J], where} \tag{2}$$

s (m) is the projectile trajectory, in this case, the thickness of the plate, and F is the extrusion force, which is calculated from the equation:

$$F = K.\pi.d.s \quad \text{[N], where} \tag{3}$$

d (mm) is the diameter of the projectile, s (mm) is the thickness of the plate and K is a parameter determining the mechanical properties of the shot plate, which is usually determined from experiment.
The Gabeaud equation was used to calculate the limiting impact velocity required to penetrate the plate:

$$v_{lim} = 125.\sqrt{\Delta}.\sqrt{\frac{d_p^2.s^2}{m.(s+h_H)-3.d_p^2.s^2}} \text{ [m.s}^{-1}\text{], where} \tag{4}$$

v_{lim} (m.s^{-1}) is the limiting velocity of the projectile if the entire kinetic impact energy E_d [J] is consumed to penetrate a plate of thickness s (dm) that can withstand a projectile calibre d_p (dm) of

IOP Conf. Series: Materials Science and Engineering 1243 (2022) 012002 doi:10.1088/1757-899X/1243/1/012002

mass *m* (kg) with a projectile head height h_H (dm). The penetration conditions are given by the quality of the shot plate Δ [MPa], into which the hardness values found by Brinell [4] are entered.
Table 6 shows the impact energies of projectiles and 20 mm FSP fragments from a 155 mm artillery shell according to NATO AEP-55 STANAG 4569.
Table 6 shows that the impact energy of the FSP 20 mm fragments is several times greater than the impact energy of the projectiles for levels 1 to 3, indicating that a greater thickness of armour plate will be required for ballistic resistance to FSP 20 mm fragments.

Table 6. Impact energy of projectiles and fragments according to STANAG 4569.

Article V.	**LEVEL 1**		
Projectile / fragments	**m [kg]**	**v [m.s^{-1}]**	**Ed [J]**
5.56mm x45 M193	0.00356	937	1563
5.56mm x45 NATO SS 109	0.004	900	1620
7.62mm x 51 NATO Ball	0.00965	833	3348
FSP 20mm	0.0538	520	7274
	LEVEL 2		
Projectile / fragments	**m [kg]**	**v [m.s^{-1}]**	**Ed [J]**
7.62mm x 39 API BZ	0.00777	695	1877
FSP 20mm	0.0538	630	10677
	LEVEL 3		
Projectile / fragments	**m [kg]**	**v [m.s^{-1}]**	**Ed [J]**
7.62mm x 54R B32 API	0.01004	854	3661
7.62mm x 51 AP (WC core)	0.0084	930	3633
FSP 20mm	0.0538	720	13945
	LEVEL 4		
Projectile / fragments	**m [kg]**	**v [m.s^{-1}]**	**Ed [J]**
14.5mm x 114 API / B32	0.064	911	26557
FSP 20mm	0.0538	960	24791
	LEVEL 5		
Projectile / fragments	**m [kg]**	**v [m.s^{-1}]**	**Ed [J]**
25mm x 137 APDS-T	0.13	1258	102867
FSP 20mm	0.0538	960	24791
	LEVEL 6		
Projectile / fragments	**m [kg]**	**v [m.s^{-1}]**	**Ed [J]**
30mm x 165 AP-T	0.403	810	132204
30mm x 173 APFSDS-T	0.23	1330	203424
FSP 20mm	0.0538	1250	42031

3 Results

The 20 mm FSP fragments, which are intended to simulate flying fragments from 155 mm calibre artillery shells, have higher impact energy for NATO AEP-55 STANAG 4569 Annex A levels 1 to 3 than the impact energies of projectiles fired from small arms.

With the manufacturer's declared ballistic resistance of ARMOX Advance plates and the impact energies of the projectiles, thicknesses were calculated that should protect vehicle crews according to NATO AEP-55 STANAG 4569 Annex A for level 1 and 2 and also against flying fragments of FSP 20 mm from 155 mm artillery shells (Table 7). In order to achieve the required ballistic resistance against flying fragments of FSP 20 mm, a plate thickness of approximately 6 mm would be required for level 1 and several ARMOX Advance plates layered on top of each other would be required for level 2 and above.

When calculating the limiting impact velocity according to the Gabeaud equation, the impact velocity for the 5.56x45 M193 projectile was calculated to be $v_{lim} = 393.5$ m.s^{-1} and for the 7.62x39 API BZ projectile the impact velocity was $v_{lim} = 510.7$ m.s^{-1}. At these impact velocities, the impact energies for the 5 mm and 7 mm ARMOX Advance plates are 275.62 J and 912.85 J, respectively, which do not match the calculated impact energies from Table 6.

Table 7. Parameter K and armour plate thickness for level 1 a 2 according to calculated work for projectiles and FSP 20 mm fragments.

Article VI.	LEVEL 1	
Projectile / fragments	**K [N.mm^{-2}]**	**s [mm]**
5.56mm x45 M193	3491.356857	5
5.56mm x45 NATO SS 109	3618.680811	5
7.62mm x 51 NATO Ball	5465.135892	5
FSP 20mm	3491.356857	5.8
	LEVEL 2	
Projectile / fragments	**K [N.mm^{-2}]**	**s [mm]**
7.62mm x 39 API BZ	1563.23301	7
FSP 20mm	1563.23301	10.5

4 Conclusion

Within the research, the resistance of ARMOX Advance armour plate was investigated according to NATO AEP-55 STANAG 4569 Annex A level 1 and 2 using the equation of impact energy of projectiles and fragments and the penetration equation, where the ballistic resistance to perforation by 5.56x45 to 7.62x39 API BZ projectiles for thicknesses of 5 mm and 7 mm was declared by the manufacturer.

The calculations show that the impact energy of flying fragments from artillery shells is higher than the impact energy of projectiles for levels 1 to 3, for levels 4 to 6 the impact energy of artillery shells is lower than the impact energy of projectiles. From the observed results it can be concluded that ARMOX Advance armour plate itself meets the ballistic resistance in the range of NATO AEP-55 STANAG 4569 Annex A standard with additional resistance to fragments from artillery shells of 155 mm calibre only for the lowest level - level 1, for higher levels ARMOX Advance can be recommended as one of the plates in layered armour. In the case of Gabeaud's equation for calculating the limiting impact velocities of projectiles, there is no use for the ballistic resistance of the ARMOX Advance armour plate.

The ballistic resistance of armour plates to small arms fire can be taken into account when armouring civilian vehicles in urban traffic where artillery fire is unlikely. For armouring of military vehicles, it is also appropriate for all levels of the NATO AEP-55 STANAG 4569 Annex A standard use the calculations for the impact energy of fragments from 155 mm artillery shells, wherein areas of armed conflict there is a high probability of artillery shells impacting on military bases and these few extra millimetres of armour on equipment can protect many human lives.

Acknowledgements
This research was funded by the Ministry of Education, Youth and Sports (MŠMT), student grant competition of the University of West Bohemia, under the project SGS 2021-025 "Application of electron beam melting for welding of advanced high strength steels".

References

[1] STANAG 4569. *Protection levels for occupants of armoured vehicles.* 2. vyd. Brusel: NATO Standardization Agency, 2012.
[2] Dimont HDX [online katalogový list]. *The steel you want between you and risk.* ©2018 [cit. 19.6.2021]. Dostupné z: https://dimonthdx.cz/wp-content/uploads/2018/05/Armox-The-steel-you-want-between-you-and-risk.pdf
[3] SSAB. *Armox Advance* [online]. Stockholm: ©2021 [cit. 19.6.2021]. Dostupné z: https://www.ssab.com/products/brands/armox/armox-advance
[4] JUŘÍČEK, Ludvík a NOVOTNÝ, Pavel. Modelování průstřelu ocelové desky malorážovou průbojnou střelou. *Vojenské zdravotnické listy.* 2004, 73(3), 97-106.

Effect of Heat Treatment Before and After Incremental Rolling on the Mechanical Properties of Titanium-Stabilized Austenitic Stainless Steel

H Jirková[1*], T Janda[1], K Opatová[1], J Říha[1], J Růžička[1], D Bublíková[1] and J Volák[1]

[1]Regional Technological Institute, University of West Bohemia, Univerzitní 8, 301 00 Pilsen, Czech Republic

[*]E-mail: hstankov@rti.zcu.cz

Abstract. Austenitic stainless steel 08Ch18N10T stabilised with titanium is used for parts of VVER nuclear power reactors. The final mechanical properties are strongly affected by the previous processing history. This is mainly due to the deformation hardening that is incorporated into the material during the production of semi-finished products. The experimental program was performed on 08Ch18N10T steel bars with a diameter of 30 mm. Incremental rolling was performed on bars that remained in the delivered condition and on bars subjected to solution annealing at 1020 °C for 60 minutes before rolling. The rolled bars were stabilised annealed at different temperatures. Then a metallographic evaluation and a tensile test at RT and 350 °C were performed. The results show that the rolling speed significantly affects the mechanical properties after stabilisation annealing. By reducing the rolling speed, it was possible in some cases to increase the yield strength up to 349 MPa. The stabilisation annealing temperature of 720 ° C with a holding time of 10 hours was evaluated as the most advantageous. In the case of higher stabilisation annealing temperatures, the lowest rolling speed, i.e. 25 rpm, had to be performed to achieve higher hot yield strength values exceeding 177 MPa.

1 Introduction

The previous processing history significantly influences the processes that take place during subsequent heat treatment. It was found that forming at lower temperatures promotes the precipitation of fine particles during subsequent annealing in 08Ch18N10T steel. Austenitic grains contain a large number of dislocations after deformation, which serve as nucleation sites for the exclusion of titanium carbonitrides. Due to their presence, the yield stress is significantly improved [1-3].

In the case of stainless austenitic steels, hot forming plays a very important role in achieving the desired mechanical properties. In low stacking fault energy materials, the main recovery process during hot forming is dynamic recrystallisation, which is influenced by a number of parameters such as temperature, the magnitude and rate of embedded deformation, and grain size [4]. Since AISI 321 is characterised by low yield strength and hardness, it is necessary to use different options to increase them [5-7]. In particular, deformation and precipitation hardening are used to increase the yield strength.

The heat treatment of this steel usually consists of two steps – solution and stabilisation annealing. During the solution annealing process, the unwanted intermetallic phases are dissolved, the structure is homogenised, and the deformation hardening introduced during the previous forming operation is

removed. During the stabilising annealing, which takes place in the temperature range from 600 to 800 °C, very fine particles of titanium carbides or carbonitrides are precipitated, which lead to an increase in strength characteristics [8]. This steel is used for a wide range of parts that are used at elevated temperatures due to its excellent corrosion properties. These include parts used in nuclear power plants, chemical reactors, automotive exhaust systems, pressure vessels, engine mufflers, piping in the food and chemical industries, etc. [1,7,9].
Due to the significant influence of forming conditions on the evolution of structure and mechanical properties, this paper focused on describing the effect of different rolling speeds and different input microstructure on the hot yield strength after stabilising annealing.

2 Experimental Program

The experimental program was carried out on rolled bars of 08Ch18N10T steel with a diameter of 30 mm (table 1). Steel 08Ch18N10T belongs to the austenitic steel group and it is stabilised with titanium. It is a material that is commonly used mainly for the production of parts for the drives of the control elements of the VVER 400 and VVER 1000 pressurised water reactors. The structure of the delivered bars was strongly deformed and the structure consisted of austenitic grains of 9.7 ± 6.7 µm, delta ferrite, titanium carbides, and carbonitrides (figure 1). The hardness value was determined to be 247 HV10. Due to the strain hardening introduced into the material during the production of the bars, the hot yield strength at 350 °C was 494 ± 2 MPa, the ultimate tensile strength was 597 ± 1 MPa and the ductility was 19 ± 0%.

Table 1. Chemical composition of the 08Ch18N10T steel bars as-delivered (wt. %)

C	Si	Mn	P	S	Cr	Ni	Mo	V	Cu	W	Ti	N2	Co
0.07	0.59	1.83	0.030	0.003	17.10	11.00	0.06	0.070	0.04	0.02	0.29	0.012	0.02

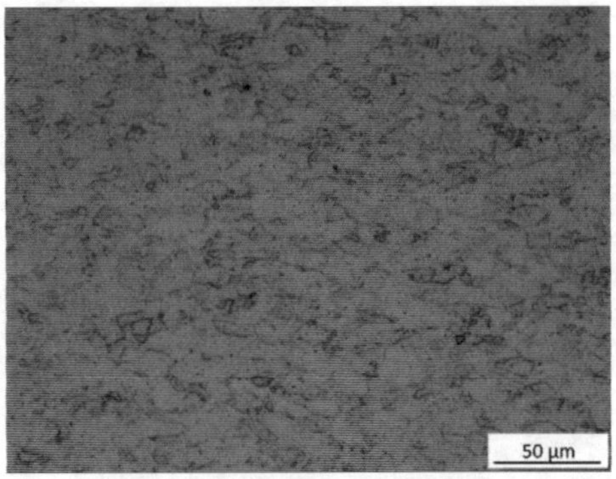

Figure 1. Deformed structure of 08Ch18N10T steel as-delivered

Incremental rolling was performed on the HDQT-R-30-12 (High Deformation Quenching and Tempering) rolling mill [10, 11]. It is an experimental device for bar rolling. The rolling is carried out by a trio of conical rollers. The rollers are offset from each other by 120° [10, 11]. The input was steel bars with a length of 850 mm and a diameter of 30 mm. The bars were heated to 970 °C by induction heating. Rolling was performed at 100, 75, and 25 rpm, followed by quenching in water. The first set of bars was rolled as-delivered (e.g. deformed). The second set of bars was first subjected to a solution annealing at 1020 °C with a soaking time of 60 minutes. Then, the water quenching followed before the rolling. A dwell time of 60 minutes was assumed to be long enough to remove the previous processing history, homogenise the structure, and for full recrystallisation of the structure. The heating was carried out without a protective atmosphere.

The bars were stabilised by annealing after rolling. The directly rolled bars were processed with the standard 720 °C/10 h, combined 680 °C/5 h + 720 °C/10 h mode, which was recommended by the experts from Škoda JS, a. s. The last mode chosen was the 800 °C procedure with a dwell time of 1 h. The use of stabilising annealing with a shorter dwell time could save energy and time during processing. A group of bars that were annealed at 1020°C prior to rolling, stabilising annealing of 720 °C/10 h, 800 °C/1 h was selected, and an extension of the dwell time at 800°C to 2.5 hours was also tested.

The microstructures of the initial state, after heat treatment and from the re-rolled bars were evaluated from the edge of the bars using an Olympus light microscope. The structure was highlighted with etchant 2:2:1 (hydrofluoric acid:glycerin:nitric acid). Austenite grain size was measured by the intersection method according to ASTM A112. The tensile test on 10 mm diameter and 50 mm length of the active part was carried out according to EN ISO 6892-1 and at 350 °C according to EN ISO 6892-2. The results were supplemented by Vickers hardness measurements at a load of 10 kg.

3. Results and Discussion

3.1. Bar rolling with subsequent stabilisation annealing

In the first step, the rolling of bars was carried out without prior solution annealing. The rolling speeds were 100, 75 and 25 rpm. To improve the mechanical properties, in particular hot yield strength, a stabilisation annealing (SA) was carried out after rolling with different parameters: 720 °C/10 h, 680 °C/5 h + 720 °C/10 h, and the more energy-efficient 800 °C/1 h annealing.

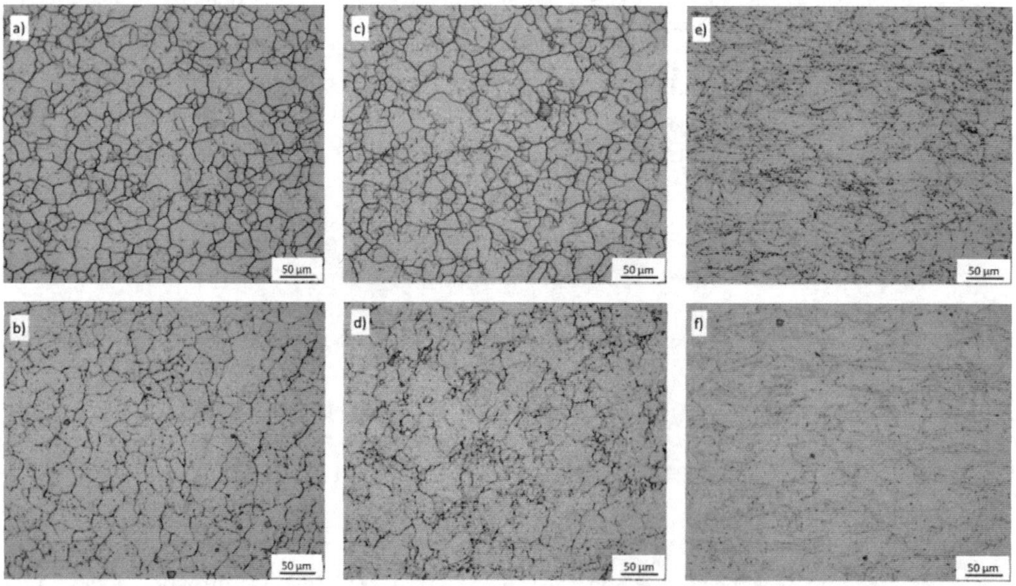

Figure 2. Microstructure after incremental rolling and stabilising annealing: a) 100 rpm – 720 °C/10 h, b) 100 rpm – 800 °C/1 h, c) 75 rpm – 720 °C/10 h, d) 75 rpm – 800 °C/1 h, e) 25 rpm – 720 °C/10 h, f) 25 rpm – 800 °C/1 h

It was found that the structure is strongly influenced by the rolling speed, and, in particular, the finish rolling temperature, which depends on it. In the case of rolling at 100 rpm, the finish rolling temperature was 1020 °C, at 75 rpm the temperature dropped to 963 °C, and at the slowest speed of 25 rpm the finish rolling temperature was only 846 °C.

When rolling at 100 rpm was performed before SA of 720 °C and 680 °C + 720 °C, a recrystallised structure with precipitates occurring along the boundaries and inside the austenitic grains was obtained (figure 2a). Using a shorter SA time at 800 °C, the character was very similar, but a lower degree of precipitation was found along the boundaries of the austenitic grains (figure 2b). The hardness value was around 170 HV10 in all cases.

When the rolling speed was decreased to 75 rpm, only partially recrystallised structures were obtained due to the reduced rolling temperature (figure 2c, d). The biggest heterogeneity in austenitic grain size was found in the case of SA at 800 °C with a soaking time of 1 h. This was also reflected in the hardness value, which was 186 HV10 in the case of SA 800 °C/1 h, while after standard SA procedures, the hardness reached 178 HV10.

At the lowest rolling speed of 25 rpm, a highly deformed structure consisting of elongated austenitic grains, precipitates along the grain boundaries, and within the grains, and titanium carbides or carbonitrides was obtained after all SA cases (figure 2e, f). Hardness ranged from 227 to 230 HV10.

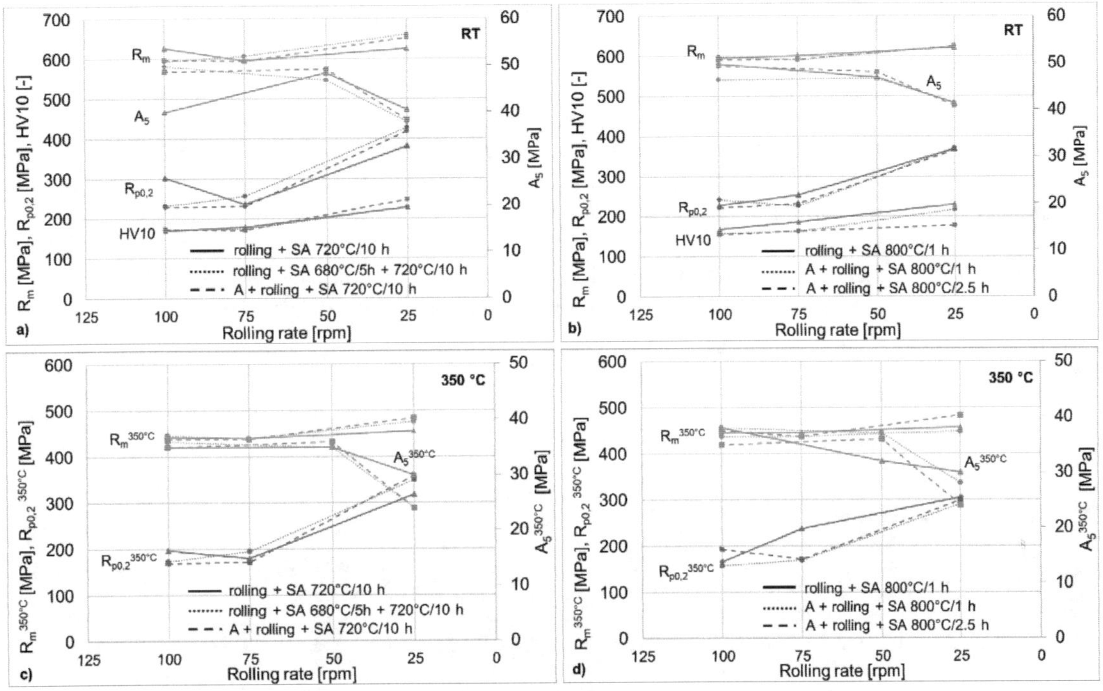

Figure 3. Mechanical properties after incremental rolling and stabilising annealing: a) RT – lower temperatures of SA, b) RT - higher temperatures of SA, c) 350°C – lower temperatures of SA, d) 350 °C - higher temperatures of SA, A – solution annealing, SA – stabilisation annealing

The effect of different post-rolling SA procedures was also evident in the mechanical properties, which were determined by tensile testing at room temperature and at 350 °C (figure 3). At RT, ultimate strength values around 600 MPa were achieved for all tested processing variants (figure 3a, 3b). A slight increase occurred at the lowest rolling speed. For the yield strength, the deformation hardening of the material and the effect of the rolling speed were significantly affected. A large increase occurred at a rolling speed of 25 rpm. After SA 720 °C/10 h, yield strength of 381± 3 MPa was achieved. For the combined SA procedure, the yield strength was even higher – 426 ± 1 MPa (figure 3a). Even with the economical variant at 800 °C, yield strength of 368 ± 1 MPa was obtained at this rolling speed (figure 3b).

In the tensile test at 350 °C, the required yield strength of over 177 MPa was obtained in all cases after using the standard SA mode of 720 °C/10 h (figure 3c). The ultimate strength was over 440 MPa with ductility of over 30 %. The yield strength was higher at 317 MPa for bars rolled at 25 rpm. In the case of the combined SA procedure, a lower yield strength was obtained at the highest rolling speeds than the procedure with only one annealing step, and these values were below the required

177 MPa. A significant increase occurred again at a rolling speed of 25 rpm when the value rose to 349 MPa, but there was a significant drop in ductility to 24 %. Reducing the dwell time to 1 h at 800 °C resulted in the lowest yield strength values of the modes studied (figure 3d). The highest yield strength of 304 ± 1 MPa was achieved at a rolling speed of 25 rpm.

3.2. Bar rolling after solution annealing with subsequent stabilisation annealing

The second part of bars with a diameter of 30 mm was annealed at 1020 °C with a dwell time of 60 minutes before rolling. The same rolling speed was used for the first set of bars. The stabilisation annealing was carried out at 720 °C with a soaking time of 10 h and at 800 °C with soaking times of 1 and 2.5 h.

In the case of a rolling speed of 100 rpm, a recrystallised austenitic structure with precipitates along the grain boundaries and within the grains was achieved in all SA cases (figure 4a-c). At a higher stabilisation annealing temperature of 800 °C, the precipitation inside the grains was more pronounced, and the envelopes formed by precipitates were not so significant along the grain borders (figure 4b, c). Titanium carbides, carbonitrides, and annealing twins were also observed in the structure. Only a very small amount of delta ferrite was detected in the structure. At a rolling speed of 25 rpm, it was found that SA at 720 °C did not lead to the removal of the deformed structure with elongated austenitic grains (figure 5a). Partial removal of the deformed structure and its recrystallisation was achieved only in the case of SA at 800 °C with a soaking time of 2.5 h (figure 5c). This is also evidenced by the gradual decrease in hardness from 248 HV10 after SA at 720 °C; 217 HV10 after SA at 800 °C with soaking time of 1h; and 177 HV10 after SA at 800 °C with soaking time of 2.5 h.

Figure 4. Microstructure of rolled bars at 100 rpm with a previous solution annealing of 1020 °C/60 minutes and SA: a) 720 °C/10 h, b) 800 °C/1 h, c) 800 °C/2.5 h

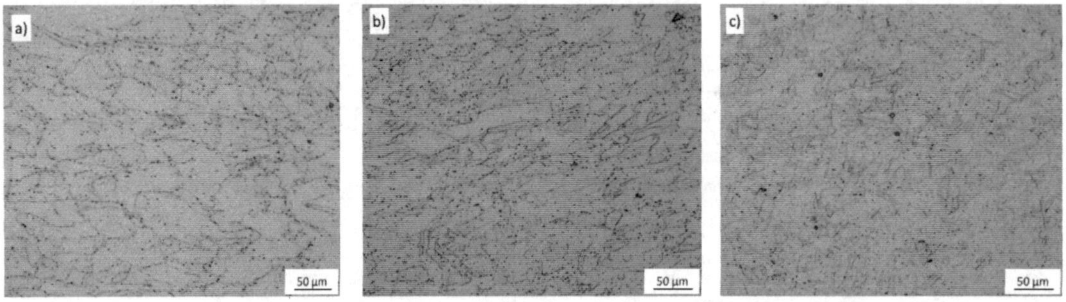

Figure 5. Microstructure of rolled bars at 25 rpm with a previous solution annealing of 1020 °C/60 minutes and SA: a) 720 °C/10 h, b) 800 °C/1 h, c) 800 °C/2.5 h

After stabilising annealing (SA), a significant increase in yield strength was achieved in all tested variants only at the slowest rolling speed, i. e. 25 rpm (figure 3a-d). In the case of the RT tensile test, the yield strength exceeded 400 MPa only in the case of SA at 720 °C (figure 3a). At the higher temperature of SA, the yield strength at the slowest rolling speed ranged from 365 to 371 MPa (figure 3b). Slower rolling speeds resulted in an increase in yield strength of 140 to 188 MPa depending on the SA mode used.

The tensile test confirmed the previous results obtained in the first part of the experiment, where the required hot yield strength of 177 MPa was only achieved at the slowest rolling speed. For SA at 720 °C, an increase in yield strength at 350 °C from 171± 0 MPa to 359 ± 1 MPa can be observed (figure 3c). After SA at 800 °C, the highest yield strength obtained was lower, only around 299 MPa (figure 3d), with yield strength values at higher rolling speeds reached 157-171 MPa.

The results show that the solution annealing before incremental rolling resulted in removing the previous processing history on the bars supplied in the cold-rolled condition. The annealed bars did not experience deformation hardening during incremental rolling at the higher rolling speeds of 100 and 75 rpm due to the higher finish rolling temperature. This resulted in lower mechanical property values at these rolling speeds in combination with SA. At a rolling speed of 25 rpm, where the finish rolling temperature was below 900 °C, the deformation hardening was significant and there was also a significant increase in the hot yield strength. These results were already comparable to the results obtained on bars without prior heat treatment.

4. Conclusion

The effect of solution annealing before incremental rolling on the properties after heat treatment was tested on bars of 08Ch18N10T. The bars were incrementally hot-rolled at 970 °C at different rolling speeds.

If the bars were not annealed before rolling, the prescribed values of mechanical properties were achieved by the standard 720 °C/10 h SA regime regardless of the previous rolling speed. To achieve the prescribed properties in the case of SA at 800 °C, rolling speeds of 75 or 25 rpm were necessary. These modes allow energy and time savings to be achieved.

In the case of rolling the pre-annealed bars at 1020 °C for 1 h, satisfactory hot yield strengths were obtained only at the slowest rolling speed of 25 rpm for all the SA temperatures tested. The highest hot yield strength of 359 ± 1 MPa was obtained after SA at 720 °C/10 h.

Acknowledgements

This research was funded by the Technology Agency of the Czech Republic under the project TJ02000274 „Determination of the principles and processes taking place during the stabilisation annealing of austenitic stainless steels used in nuclear power“ and by the Ministry of Education, Youth and Sports (MŠMT), student grant competition of the University of West Bohemia, under the project SGS 2021-025 "Application of electron beam melting for welding of advanced high strength steels".

References

[1] Nkhoma R K C, Siyasiya Ch W, Stumpf W E 2014 Hot workability of AISI 321 and AISI 304 austenitic stainless steels *Journal of Alloys and Compounds* 595 103-112

[2] Ghazani M S, Eghbali 2018 Characterisation of the hot deformation microstructure of AISI 321 austenitic stainless steel *Materials Science and Engineering: A* 730(April) 380–390

[3] Petrman I 1980 Proc. Material problems of light water reactors (Vsetín), Some problems of production of high-alloy steels for components of nuclear power plants type VVER 440, Praha : Dům techniky ČSVTS p 23-27

[4] Mirzadeh H, Parsa M H, Ohadi D 2013 Hot deformation behavior of austenitic stainless steel for a wide range of initial grain size *Materials Science and Engineering A.* 569 54–60

[5] Tiamiyu, A A et al. 2017 Development of Ultra-Fine-Grained Structure in AISI 321 Austenitic Stainless Steel *Metallurgical and Materials Transactions A: Physical Metallurgy and Materials Science* 48(12) 5990–6012

[6] Tiamiyu A A, Odeshi A G, Szpunar J A 2017 Multiple strengthening sources and adiabatic shear banding during high strain-rate deformation of AISI 321 austenitic stainless steel: Effects of

IOP Conf. Series: Materials Science and Engineering 1243 (2022) 012003 doi:10.1088/1757-899X/1243/1/012003

grain size and strain rate *Materials Science and Engineering A* 711(November 2017) 233–249

[7] XU, D et al. 2018 Effect of cold deformation on microstructures and mechanical properties of austenitic stainless steel *Metals* 8(7) 1–14

[8] Moura V, Kina A Y, Taveres S S M, Lima L D and Mainier F B 2008 Influence of stabilisation heat treatments on microstructure, hardness and intergranular corrosion resistance of the AISI 321 stainless steel *Journal of Materials Science* 43(2) 536–540

[9] Guan K, Xu X, Xu H, Wang Z 2005 Effect of aging at 700°C on precipitation and toughness of AISI 321 and AISI 347 austenitic stainless steel welds *Nuclear Engineering and Design* 235(23) 2485-2494

[10] Peković M. et al. 2018 Termomechanické zpracování ocelí s použitím zařízení pro inkrementální tváření tyčí HDQT-R 30-12 (Thermomechanical treatment of steels with using the device of incremental bars forming HDQT-R 30-12) *Kovárenství* 65 26-29

[11] Jirkova H, et al 2021 Influence of rolling speed during hot forming using HDQT-R 30-12 machine on the properties of stainless steel 08Ch18N10T, IOP Conf. Ser.: Mater. Sci. Eng. 1178 012025

Effect of chemical-heat treatment on the properties of titanium

D Bricín[1, 2*], L Kratochvílová[1], R Leták[1], L Vojta[1], K Nová[1], Z Špirit[1, 2]

[1]Department of Material Science and Technology, University of West Bohemia in Pilsen, Univerzitní 8, 301 00, Pilsen, Czech Republic

[2]Centrum výzkumu Rez s.r.o., Morseova 1245/6, 301 00, Pilsen, Czech Republic

*Email: bricda@kmm.zcu.cz

Abstract. The aim of this study is to evaluate the properties and microstructure of pure titanium after its chemical heat treatment (CHT) by boriding and nitriding. Nitriding and boriding were both done in powder backfill for ten hours. The temperature of the nitriding was 550 °C. The temperature of the boriding was 1000 °C. Both processes led to the creation of a thin surface layer of nitrides and borides which affected wear resistance and also they affected corrosion resistance of the surface of the analyzed samples.

Key words: Boriding. Nitriding. Titanium. Wear resistance. Corrosion resistance

1. Introduction

Titanium and its alloys are construction materials which are used in a wide range of applications from the automotive industry, through aviation to biomedicine [1-2]. In biomedicine they are used for knee joints, dental implants or fracture fixation screws due to their corrosion resistance and biocompatibility [2]. However, a certain disadvantage of the use of titanium and its alloys may be their lower resistance to wear, which is often associated with, for example, imperfections in its surface diffusion layer TiO_2 [3]. One of the ways to increase the wear resistance of titanium is by using suitable surface treatments. The aim of our study was to verify how chemical-thermal treatment of titanium by nitriding and boriding affects its properties. The experiments focused on the evaluation of the changes in the samples of wear resistance, corrosion resistance and microstructure caused by the chemical-heat treatment.

2. Experimental methodology

Pure titanium in the form of a sheet with a nominal thickness of 4 mm was used as the experimental material. Samples measuring 100 mm x 40 mm x 4 mm were created from this sheet. Before further processing, the surfaces of the samples were unified by grinding, using an MD-Piano 1200 grinding wheel in the last step of the process. Then, the samples were divided into four groups, three of which underwent chemical-thermal treatment processing. The parameters of the chemical-heat treatment are summarized in Table 1.

Table 1. Parameters of CHT processes

Samples series	Process type	Medium	Temperature (°C)	Time (h.)
1	-	-	-	-
2	Boriding	powder backfill	1 000	10
3	Nitriding	powder backfill	550	10
4	Combination of boriding and nitriding	powder backfill	1 000/550	5/5

The evaluation of the changes caused by CHT focused mainly on the evaluation of the surface of the samples. Metallographic analysis using a Carl-Zeiss Observer Z1m light microscope and an Olympus LEXT-OLS5100 confocal scanning microscope were used to evaluate the thickness of the diffusion layers formed and the changes caused by CHT processes in the microstructure of the samples. Cross-sections of some of the samples were prepared and etched using 'Kroll' etchant. The change in grain size was determined using the intersection method, taking into account the standard ISO 643 [4]. The Olympus LEXT-OLS5100 confocal scanning microscope was also used to analyze the topography of the surface, to evaluate changes in surface roughness, and to measure tribological traces that were created to determine differences in the surface abrasion resistance of samples after CHT. The change in the abrasion resistance of the sample surface was measured taking into account the ASTM G99 standard using the Ball on Disk test, where a ball-shaped indenter is pressed into the rotating surface of the analyzed sample. In this experiment, a sapphire ball with a radius of 2.5 mm was used. The force with which the indenter was pressed into the surface of the samples was set to 5 N. The number of repetition cycles was set to 1,000. The wear coefficient was then calculated according to Equation 1-2 [5]. The calculated data, which include the calculation of grain size, diffusion layer thickness, surface roughness and wear coefficient, are supplemented by the standard deviation of the measurements.

$$V = \frac{\pi \cdot (wear\ track\ radius\ [mm]) \cdot (track\ width\ [mm^3])}{6(sphere\ radius\ [mm])} (mm^3) \tag{1}$$

$$W = \frac{V}{L \cdot s}\left(bn \frac{mm^3}{N \cdot m}\right) \tag{2}$$

Where:
V… removal of material (mm^3)
s ... Total length of the indenter movement (m)
L… Applied force Fn (N)

The cohesive-adhesive behavior of the formed diffusion layers was evaluated using the Mercedes Test. In this test, an indenter, in this case a diamond cone with an apex angle of 120°, is pressed into the surface of the sample with a force of 150 kgf. The surroundings of the formed craters were then captured using a PHILIPS XL30ESEM scanning electron microscope. The metallographic images were then compared. The change in corrosion resistance of the samples was analyzed using an SP-150 potentiostat. The measured data were processed using EC-Lab laboratory software. The corrosive environment was chosen with regard to the possible use of these materials in biological implants. The corrosive mediums were chosen according to reference 6. The physiological solution (aqueous NaCl solution) was further combined with hydrogen peroxide and hydrogen peroxide and tartaric acid.

3. Results of the experiment

Titanium and its alloys gradually lose their strength when exposed to temperatures higher than 500 °C [1]. This loss in the strength is associated with the coarsening of the grains of its matrix. Thus, the most significant coarsening occurred in samples that were processed by boriding, or a combination of boriding and nitriding, see Fig. 1.

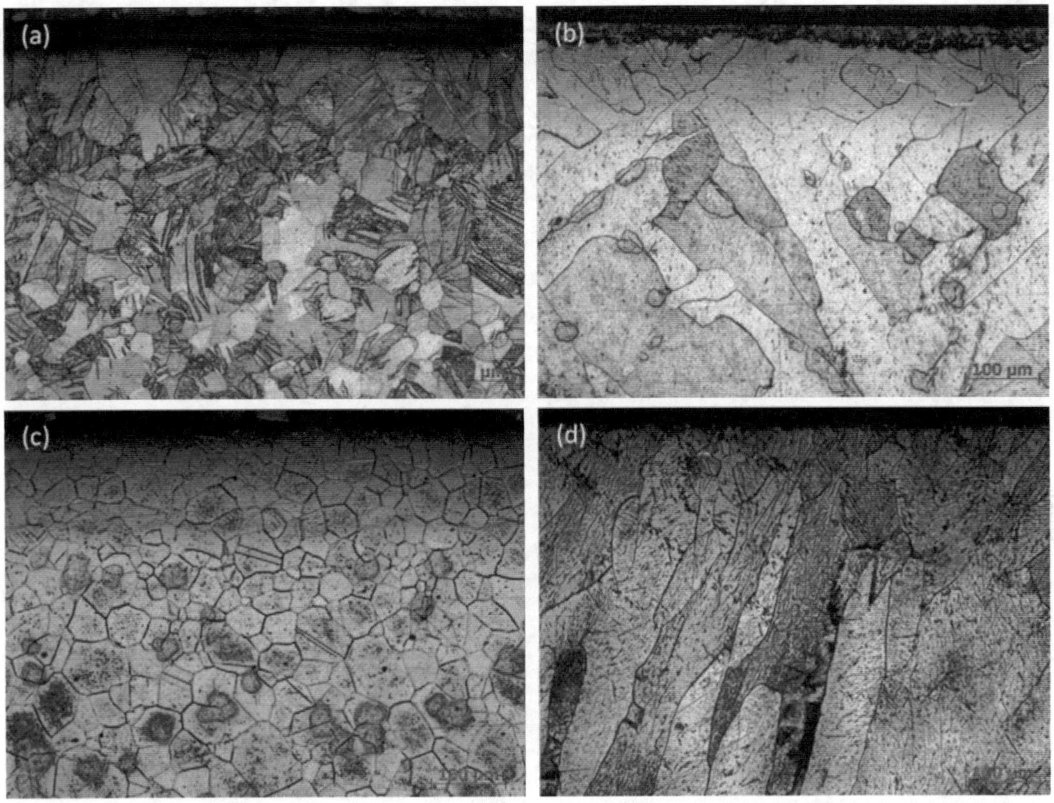

Figure 1. Microstructure of samples after CHT (a) - initial structure, (b) - Structure after boriding, (c) - Structure after nitriding, (d) - Structure after combined CHT

The original grain size of the matrix changed from 35±4.5 µm to 67±15.8 µm for the borided samples, to 35±4.1 µm for the nitrided samples, and to 68±25 µm for the samples that underwent the combined CHT process. In addition to the grain size of the heat-treated titanium matrix, the processing time also affected the thickness of the diffusion layers. The thickness of the titanium nitride diffusion layer was about 0.95±0.2 µm, copying the traces formed when grinding the sample surfaces (see Fig. 2b). In contrast, during boriding, a layer thickness of 36.3±2.6 µm was achieved, with a significantly different morphology compared to the samples after nitriding, see Fig. 2a. In the case of the combined CHT process, a combined diffusion layer was formed. This layer consisted partly of a boride layer 25.7±1.5 µm thick and partly of a nitrided layer 7.6±0.8 µm thick. There was then a sharp interface between the individual layers. Because the mutual adhesion of the layers was low, the outer nitride layer peeled off significantly during the Mercedes test. In contrast, the monolayers formed during the nitriding and boriding process showed, in particular, a susceptibility to cohesive failure, but not adhesive failure, see Fig. 2.

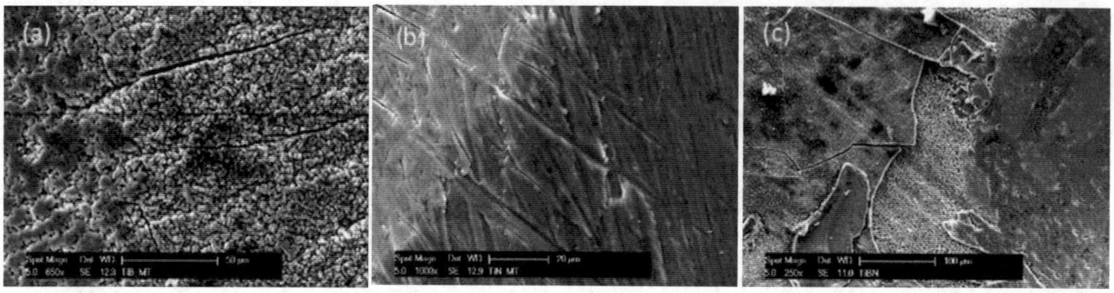

Figure 2. Comparison of the crater edges created during the Mercedes test. (a) - cohesive failure of the borided sample, (b) - cohesive failure of the nitrided sample, (c) - cohesive-adhesive failure of the sample after the combined CHT process

The formation of diffusion layers was manifested by a change in the surface roughness of the samples. Its most significant increase occurred in the borided samples and samples that underwent the combined CHT process, see Fig. 3.

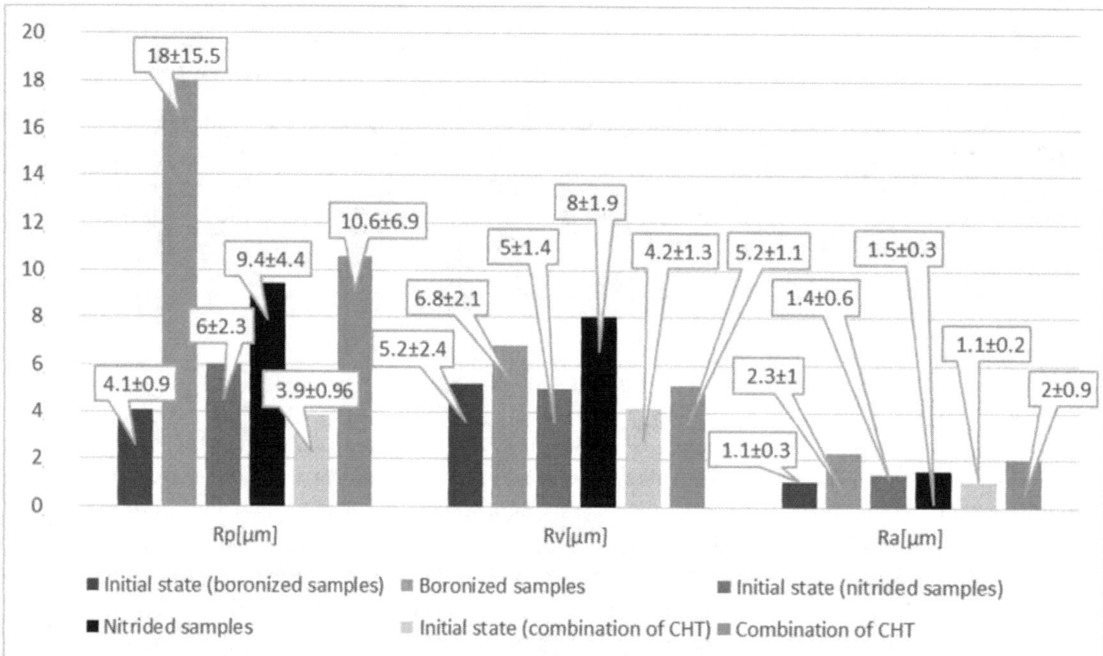

Figure 3. Comparison of selected surface roughness values measured on samples before and after CHT

However, the resulting roughness of the samples after CHT was partially affected by the fact that powder particles adhere to the surface of the samples during CHT. These must then be removed from the surface, which can be difficult. These particles affected the surface roughness of the samples after CHT and together with the cohesive-adhesive behavior of the diffusion layers significantly affected its resistance to tribological wear, see Fig. 4. The graph shows that the samples reached the highest values of wear resistance after the nitriding process, and using the combined CHT method.

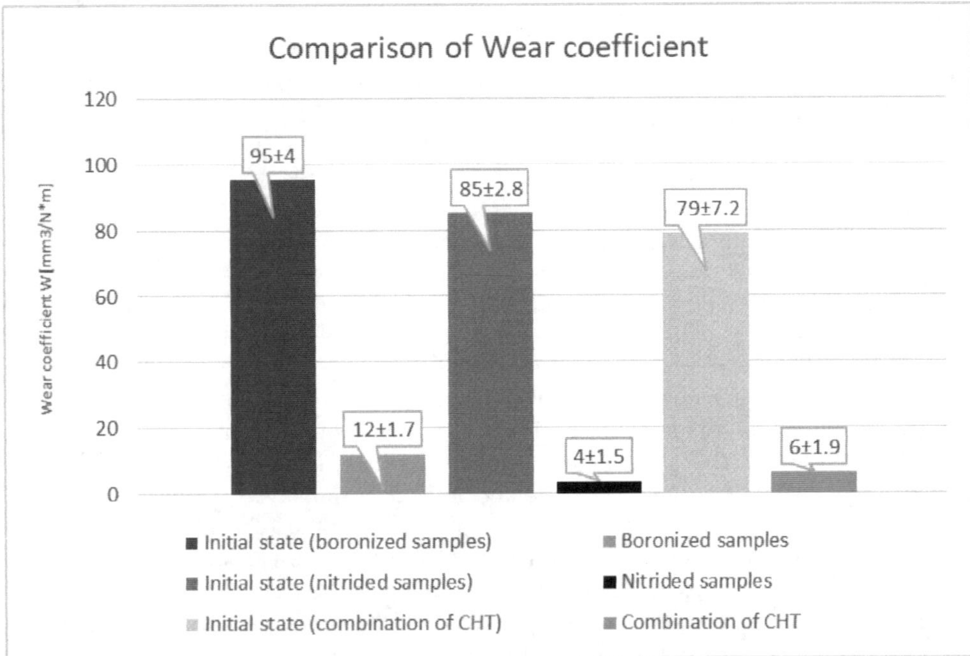

Figure 4. Comparison of the wear coefficient measured on samples before and after the CHT process

Table 2 summarizes the results of the corrosion resistance measurements of samples that have undergone CHT. The data in the table show that the nitrided samples and samples in which no CHT process was performed had the highest corrosion resistance. The samples after boriding showed the lowest corrosion resistance.

Table 2. Comparison of corrosion resistance of samples before and after CHT in different corrosion environments.

Sample type	Area (cm^2)	Solution	Rp ($Ohm \cdot cm^2$)	Vcor (μm/year)
Initial state		0.9 % NaCl (FR)	439 666.2	0.76
		FR + H_2O_2	131 210.1	2.29
		FR + NaOH, tartaric acid (pH=2.5)	369 067.6	0.92
Borided		0.9 % NaCl (FR)	12 151.1	21.11
		FR + H_2O_2	5 860.5	29.35
		FR + NaOH, tartaric acid (pH=2.5)	31 258.2	7.68
Nitrided	1.77	0.9 % NaCl (FR)	1 527 359.6	0.13
		FR + H_2O_2	1 568 991.7	0.19
		FR + NaOH, tartaric acid (pH=2.5)	1 490 044.4	0.23
Combination of CHT		0.9 % NaCl (FR)	44 250.0	6.43
		FR + H_2O_2	13 276.8	2.26
		FR + NaOH, , tartaric acid (pH=2.5)	78 844.7	2.91

4. Conclusion

The change in grain size of the titanium sample matrix was primarily temperature-dependent, and secondarily time-dependent. No significant change in matrix grain size was observed for the nitrided samples. The borided samples and samples that underwent combined CHT then had an almost twofold increase in matrix grain size. It is known that the driving force of grain growth is the energy of the grain boundaries [7]. As the grains of the material become coarser, the number of their boundaries decreases and thus this energy is reduced. As the energy of the grain boundaries decreases, the rate of coarsening of the grains gradually decreases, which gradually leads to its termination. Subsequent heat treatment at lower temperatures, such as when nitriding after the boriding process, no longer causes further grain coarsening of the material, because the stored energy will be too low. The formation of diffusion layers in all cases resulted in an increase in the surface roughness of the samples. The surfaces of the samples after boriding showed the highest increase in the size of the outlets and grooves. This fact can be advantageous for increasing its biocompatibility. As mentioned in reference 8, higher surface roughness can lead to easier adhesion of tissues to the implant surface. The thickness of the diffusion layers formed ranged, depending on the parameters of the CHT process, from about one micrometer to about 38 micrometers. The thickness of the diffusion layers largely depended on the mutual solubility of the diffusing elements in the crystalline layer of the substrate and the parameters of the CHT process. The diffusion layers further increased the resistance of the substrate to tribological wear. In terms of adhesion, however, the combined layer showed the worst cohesion, which is unsuitable when considering it for possible use in the production of implants. Nitriding contributed to increasing the corrosion resistance of the titanium surface. In contrast, boriding or its combination with nitriding reduced the corrosion resistance. However, the results show a positive effect of nitrogen on the corrosion resistance of the substrate.

Acknowledgments
The creation of this study was supported by the project SGS-2021-030 "Development of new materials, application of modern methods of their processing, ecological production, welding and testing"

References

[1] Takahashi, K., Mori, K. and Takebe, H., 2020. Application of Titanium and its Alloys for Automobile Parts. MATEC Web of Conferences, 321, p.02003.A reference
[2] Elias, C., Lima, J., Valiev, R. and Meyers, M., 2008. Biomedical applications of titanium and its alloys. JOM, 60(3), pp.46-49
[3] Long, M. and Rack, H., 2001. Friction and surface behavior of selected titanium alloys during reciprocating-sliding motion. Wear, 249(1-2), pp.157-167
[4] ISO 643. (2003). Steels - Micrographic determination of the apparent grain size. Available from:https://www.iso.org/obp/ui/#iso:std:iso:643:ed-4:v2:en
[5] ASTM G99-17. (2017). Standard Test Method for Wear Testing with a Pin-on-Disk Apparatus. ASTM International, West Conshohocken, PA. www.astm.org
[6] Bachmayer, M. Kostní implantáty na bázi železa. Brno: Vysoké učení technické v Brně, Fakulta elektrotechniky a komunikačních technologií, 2015. 80 s. Vedoucí bakalářské práce doc. Ing. Marie Sedlaříková, CSc.
[7] Gil, F. and Planell, J., 2000. Behaviour of normal grain growth kinetics in single phase titanium and titanium alloys. Materials Science and Engineering: A, 283(1-2), pp.17-24.
[8] Major, S., Cyrus, P. and Hubálovská, M., 2017. The Influence of Surface Roughness on Biocompatibility and Fatigue Life of Titanium Based Alloys. IOP Conference Series: Materials Science and Engineering, 175, p.012053.

IOP Conf. Series: Materials Science and Engineering 1243 (2022) 012005 doi:10.1088/1757-899X/1243/1/012005

Operational verification of reuse of returnable zinc alloy ZP0410 in the production of thick-walled castings

K Gryc[1*], L Socha[1], R Kubeš[2], V Sochacký[2], K Koza[1] and J Trobl[2]

[1]Institute of Technology and Business in České Budějovice, Okružní 517/10, 370 01 České Budějovice, Czech Republic

[2]GD Druckguss s.r.o., Radomilická 1244, 389 01 Vodňany, Czech Republic

[*]Email: gryc@mail.vstecb.cz

Abstract. Nowadays, more profound implementation of circular economy elements in individual production processes is ever more emphasized. In an effort to reduce ZL0410 primary raw material consumption to produce castings made of the ZP0410 alloy, GD Druckguss s.r.o. engaged in the development of a recycling technology for returnable materials based on ZP0410. A long-term series of test melts focused on monitoring of the pre-melting process was carried out from 2019 to 2021. Samples were taken regularly from individual technological nodes of thick-walled casting production. The samples of input materials such as the primary alloy ZL0410, returnable material ZP0410, and also solidified melted mass of ZP0410 and ZP0410 castings, were analysed using optical emission spectrometry in the Q4 TASMAN system to determine compliance with chemical composition according to ČSN EN 12844. The standard reference material SRM 630 was used in the developed measurement methodology for interim checks and for corrections of the system settings. It was demonstrated that the chosen ratio of raw material (ZL0410) and returnable material (ZP0410) of 2:1 (wt.) has no negative impact on compliance with chemical composition of castings made of the ZP0410 alloy according to ČSN EN 12844.

1 Introduction

In 2020, world mine production of zinc reached 12 million tons [1]. Fifty percent of zinc production is used for steel galvanization, and 15 % is used in the form of alloys for casting production [2]. Casting production is associated with a large amount of waste in the form of inlet systems and feeds; this waste represents up to 55 % of the gross casting mass before working [3]. An increased recycling rate of this waste would be desirable in terms of environmental protection; at the same time, it would improve production sustainability [4]. Not much of relevant information is available on the currently used recycling rates. The study [3] mentions the ratio of 1:1 (wt.) of primary raw material and recycled waste, suggesting increasing this ratio to 2:3 (wt.), noting, however, the quantity of sub-standard pieces might grow.

In 2020, GD Druckguss, a company engaged in the production of high-quality thick-walled zinc alloy castings, used 1057 tons of the primary raw material ZL0410; recycled waste – inlet systems from previous melts – was added in the ratio of 2:1 (wt.). This was performed in order to increase the effectiveness of the production process. Besides a direct positive economic effect of reduced production costs and costs of transporting returnable material for external refining, energy demands of the production process were also decreased. Additionally, recycling ensures that the strategic material stays in the economic cycle in the Czech Republic.

The quality of the primary raw material ZL0410 is defined in ČSN EN 1774 [5], and the quality of ZP0410 castings is defined in ČSN EN 12844 [6]. These standards determine, among other things, chemical composition of the molten mass and castings, which is an important parameter in the production and in the final product.

The paper studies the effect of adding recycled waste to the primary raw material ZL0410 on chemical composition of the alloy in the course of casting production. The objective is to demonstrate that zinc waste recycling has no negative impact on chemical composition of the alloy and thus that it can be applied on the industrial scale.

2 Experimental conditions

In the production of high-quality thick-walled zinc alloy castings, input materials comprising primary raw material ZL0410 (two units) and recycled waste (one unit) are first melted in the pre-melting furnace MELTEC ZTF 700 B. Subsequently, the melt is transported to individual casting machines in the pre-heated mobile furnace MELTEC ZTF 650 (Figure 1). The melt is pumped from the mobile furnace to holding furnaces of individual casting machines for pressure casting. Series of samples were taken from the above-mentioned aggregates in five production cycles in order to monitor the quality of melt in the technological flow. The samples were taken into prepared samplers. Furthermore, samples of recycled waste and finished castings were analysed. The composition of melt in the pre-melting furnace MELTEC ZTF 700 B was monitored from 2019 to 2021 in order to determine any potential long-term trends. Again, the samples were taken into prepared samplers.

Figure 1. Pre-melting furnace MELTEC ZTF 700 B with a recycled waste conveyor (left), with the pre-heated mobile furnace MELTEC ZTF 650 (right).

Chemical composition of the samples was analysed using optical emission spectrometry in the Q4 Tasman system. Before measurements, the analysed surface of the samples was ground using a disk grinder to remove any superficial inhomogeneities and contaminations.

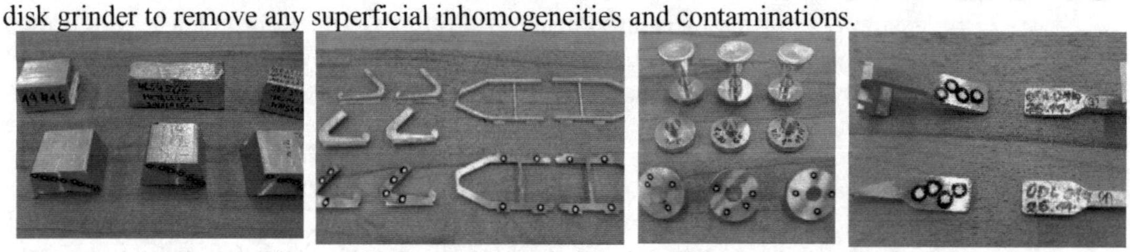

a) Primary raw material b) Returnable material c) Melt d) Casting

Figure 2. Examples of samples from operational melts after their analysis using Q4 TASMAN.

IOP Conf. Series: Materials Science and Engineering 1243 (2022) 012005 doi:10.1088/1757-899X/1243/1/012005

The standard reference material SRM 630 was used in the analysis for interim quality checks and to make sure that correct results would be obtained. Figure 2 shows examples of samples after their analysis in the optical emission spectrometer. Furthermore, selected samples underwent a review analysis using the Q8 Magellan instrument to verify the quality of the results.

3 Results and discussion

Table 1 provides standardized composition of the zinc alloy for the primary raw material ZL0410 (ČSN EN 1774) and for the castings ZP0410 (ČSN EN 12844).

Table 1. Zinc alloy composition according to the standards.

	Element content (wt. %)			
	ČSN EN 1774		ČSN EN 12844	
	min	max	min	max
Al	3.800	4.200	3.700	4.300
Cu	0.700	1.100	0.700	1.200
Mg	0.035	0.060	0.025	0.060
Pb	-	0.003	-	0.005
Sn	-	0.001	-	0.002
Si	-	0.020	-	0.030
Cd	-	0.003	-	0.005
Ni	-	0.001	-	0.020
Fe	-	0.020	-	0.050

Figures 3–8 present the determined content of Al, Cu, Mg, Pb, Sn; and Si. Al, Cu and Mg are alloying element. Increased content of Pb, Sn and Si is considered harmful. Values of other harmful elements (Cd, Ni, Fe) were also monitored; however, their contents always complied with the standard requirements for castings (ČSN EN 12844), and therefore they are not further discussed in this paper. For each element, one graph shows the content in individual steps of the production process where SCRAP means recycled waste; FURNACE is the pre-melting furnace MELTEC ZTF 700 B; TRANS is the mobile furnace MELTEC ZTF 650; MACH10 and MACH14 are holding furnaces of the casting machines; and CAST10 and CAST14 are castings from the given casting machines. The green area indicates the range of the ČSN EN 12844 standard. Blue points comply with the standard, while red points do not; the mean value is indicated in black. The second graph for each element provides the content of the given element in the pre-melting furnace MELTEC ZTF 700 B in the period of 2019–2021.

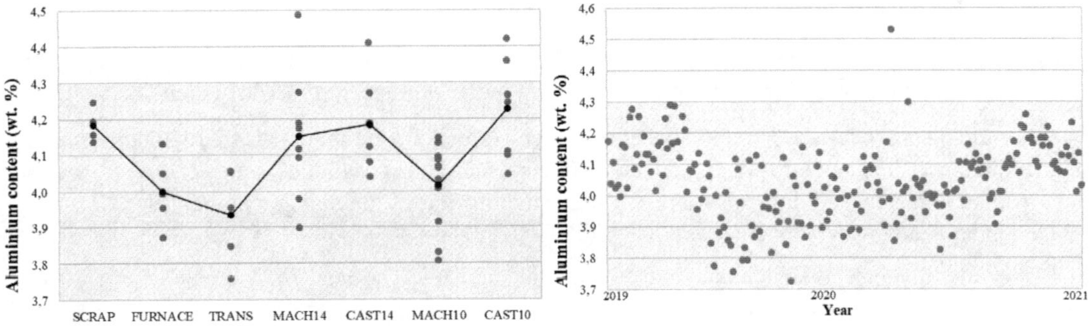

Figure 3. Aluminium content in individual steps of the production process (left); long-term monitoring of aluminium content in the pre-melting furnace (right).

IOP Conf. Series: Materials Science and Engineering 1243 (2022) 012005 doi:10.1088/1757-899X/1243/1/012005

Figure 3 shows the aluminium content in the production process. Based on the standard, aluminium content should be 3.7–4.3 wt. %. Clearly, the standard requirements were complied with in most cases and were exceeded only once in machine 14 and in one casting from this machine by 0.186 wt. % and 0.109 wt. %, respectively. Additionally, two castings from machine 10 exceeded the standard requirements by 0.058 wt. % and 0.120 wt. %, respectively. As follows from long-term monitoring of aluminium content in FURNACE, the standard requirements were complied with in all cases except one, where they were exceeded by 0.23 wt. %.

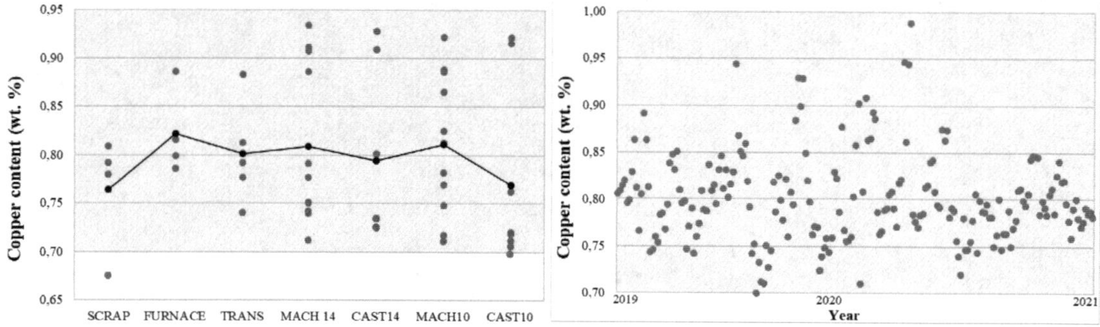

Figure 4. Copper content in individual steps of the production process (left); long-term monitoring of copper content in the pre-melting furnace (right).

Figure 4 shows the copper content in the production process. Based on the standard, copper content should be 0.7–1.2 wt. %. Clearly, the standard requirements were complied with in most cases; only one sample of recycled waste contained less copper than required, by 0.025 wt. % and one casting from the casting machine 10 contained less copper than required, by 0.002 wt. %. As follows from long-term monitoring of copper content in FURNACE, the standard requirements were complied with.

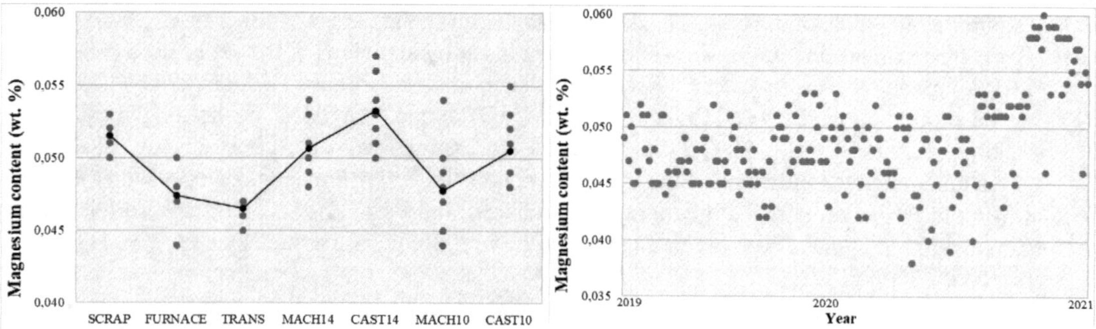

Figure 5. Magnesium content in individual steps of the production process (left); long-term monitoring of magnesium content in the pre-melting furnace (right).

Figure 5 shows the observed magnesium content values. Based on the standard, magnesium content should be 0.025–0.060 wt. %. Clearly, the standard requirements were complied with in all production steps. As follows from long-term monitoring of magnesium content in FURNACE, the standard requirements were complied with.

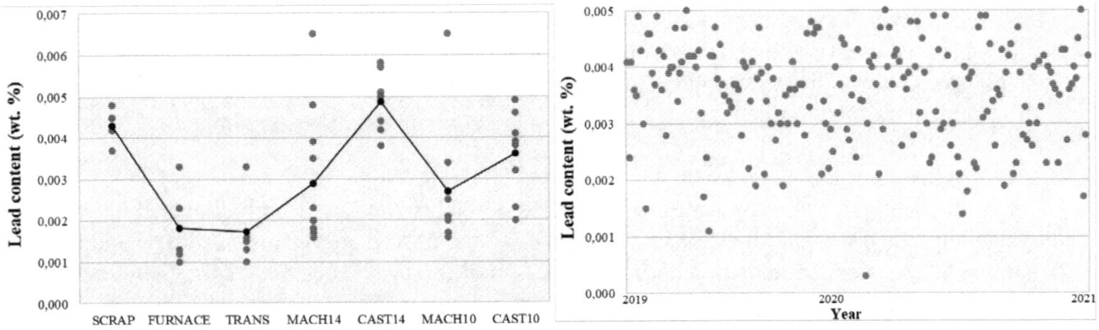

Figure 6. Lead content in individual steps of the production process (left); long-term monitoring of lead content in the pre-melting furnace (right).

Figure 6 provides the determined lead content values. Based on the standard, the maximum lead content is 0.005 wt. %. Clearly, the standard requirements were complied with in most cases; they were exceeded only once in machine 14 and once in machine 10, by 0.0015 wt. % in both cases. Furthermore, 3 castings from machine 14 exceeded the standard requirements by 0.0001 wt. %, 0.0007 wt. % and 0.0008 wt. %, respectively. As follows from long-term monitoring of lead content in FURNACE, the standard requirements were complied with.

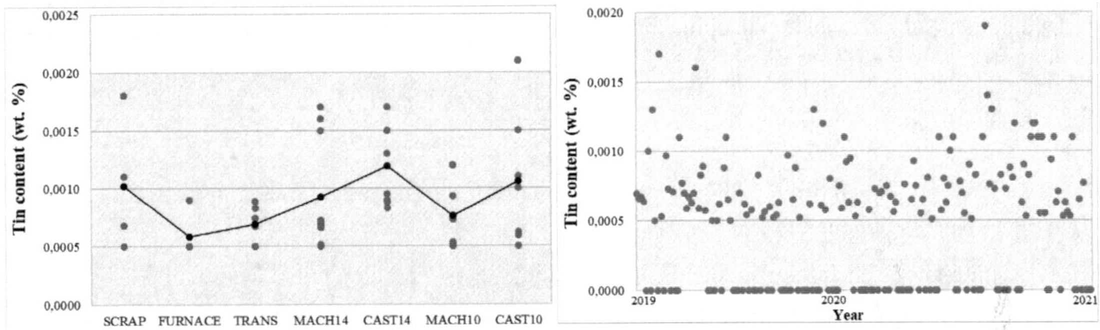

Figure 7. Tin content in individual steps of the production process (left); long-term monitoring of tin content in the pre-melting furnace (right).

Figure 7 shows the tin content in the production process. Based on the standard, the maximum tin content is 0.002 wt. %. Clearly, the standard requirements were complied with in most cases; they were exceeded only in one casting from machine 10 by 0.0001 wt. %. As follows from long-term monitoring of tin content in FURNACE, the standard requirements were complied with.

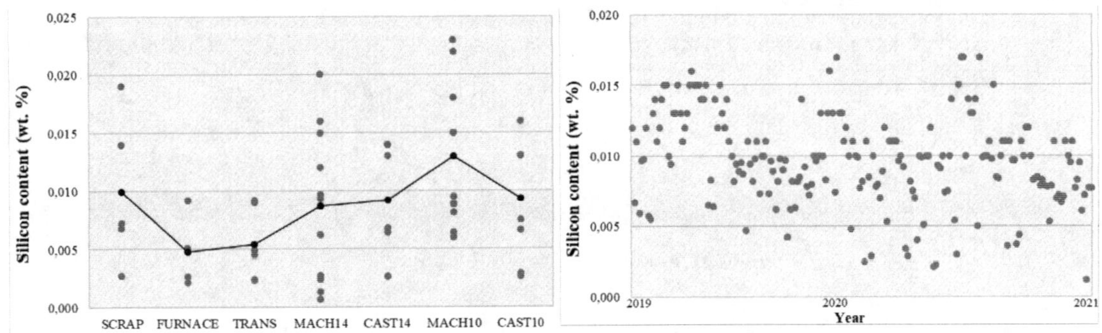

Figure 8. Silicon content in individual steps of the production process (left); long-term monitoring of silicon content in the pre-melting furnace (right).

IOP Conf. Series: Materials Science and Engineering 1243 (2022) 012005 doi:10.1088/1757-899X/1243/1/012005

Figure 8 shows the silicon content in the production process. Based on the standard, the maximum silicon content is 0.03 wt. %. Clearly, the standard requirements were complied with in all production steps. As follows from long-term monitoring of silicon content in FURNACE, the standard requirements were complied with.

4 Conclusion

The possibility of zinc waste recycling was verified in the production of high-quality thick-walled zinc alloy castings. Zinc waste of the foundry was added to the primary raw material ZL0410 in the ratio of 2:1 (wt.), and chemical composition of the alloy was monitored. The chemical composition was verified in individual steps of the 5 cycles of the production process including the recycled waste and finished castings. Additionally, chemical composition of the melt in the pre-melting furnace has been monitored since 2019 to capture any long-term trends. As shown by the measurements, the chemical composition complied with the standard requirements for castings almost in all cases. The requirements were exceeded in sporadic cases; however, these cases were rare and had no substantial impact on overall long-term quality of the castings. Thus, it was demonstrated that the pre-set ratio of the recycled waste (one unit) and primary raw material (two units) could be applied in the production of high-quality thick-walled zinc alloy castings.

Acknowledgements
The paper was prepared in the scope of the EPSILON programme, project TH04020055: Research and development of zinc waste recycling technology in the production of high-quality zinc alloy castings, supported from the resources of the Technology Agency of the Czech Republic.

References

[1] U.S. Geological Survey 2021 Mineral commodity summaries 2021 https://doi.org/10.3133/mcs2021

[2] Antrekowitsch J, Steinlechner S, Unger A, Rösler G, Pichler Ch and Rumpold R 2014 *Handbook of Recycling*, ed Worrell E and Reuter M A (Amsterdam: Elsevier), ISBN 9780123964595

[3] Tan R B H and Khoo H H 2005 Zinc casting and recycling *Int. J. Life Cycle Assessment* **10**, pp 211-218, https://doi.org/10.1065/lca2004.08.172

[4] Matinde E, Simate G S and Ndlovu S 2018 Mining and metallurgical wastes: a review of recycling and re-use practises *J. S. Afr. Inst. Min. Metall.* **118**, pp 825-844, http://dx.doi.org/10.17159/2411-9717/2018/v118n8a5

[5] ČSN EN 1774. Zinc and zinc alloys – Alloys for foundry purposes – Ingot and Liquid. 2000

[6] ČSN EN 12844. Zinc and zinc alloys – Castings – Specifications. 2000

IOP Conf. Series: Materials Science and Engineering 1243 (2022) 012006 doi:10.1088/1757-899X/1243/1/012006

Optimization of ductile iron casting production based on numerical simulation

L Socha[1*], K Gryc[1], J Sviželová[1], M Chmiel[2], M Brathová[2], S Filipek[1,3]

[1]Institute of Technology and Business in České Budějovice, České Budějovice, Czech Republic, EU
[2]KOVOSVIT MAS Foundry, a.s., Sezimovo Ústí, Czech Republic, EU
[3]University of West Bohemia, Plzeň, Czech Republic, EU

[*]E-mail: socha@mail.vstecb.cz

Abstract. The research involved numerical simulations in MAGMASOFT® focused on optimising the casting process and solidification of a cast piece type intended for the shipping industry. The paper presents the results of the numerical simulations together with a proposal of changes in the production technology aimed at increasing the casting quality. No serious deficiencies of the proposed technology were found based on the simulation results. However, as regards optimisation of the filling process, it was recommended to adjust the gates shape to prevent metal splashing. It is recommended to optimise filling of the castings in order to achieve a structure free of internal defects, and also to slightly modify the shape in order to eliminate the risk of porosity and cracking.

1. Introduction

In general, cast iron is a widely used material in the industry. For example, almost 76 mil. tons of cast iron were produced in 2019; of this, 23.6 mil. tons were represented by ductile cast iron (for comparison, the worldwide steel production was 10 mil. tons) [1]. The largest portion of the worldwide cast iron production is applied in the automotive industry. However, its applications can be encountered in many other industries [2]. Moreover, this material becomes ever more competitive to steel products given constant innovations in the casting process of cast pieces made of ductile cast irons [3].

Major factors that affect mechanical properties of cast iron include graphite morphology, composition of the metal matrix and casting defects [4, 5]. Graphite may be present in cast irons in several modifications – lamellar (GJL), spheroidal (GJS), vermicular (GJV) and malleable (GJM) [6]. In terms of mechanical properties, spheroidal graphite, contained in ductile cast iron, is the most favourable modification as it eliminates notch effects of graphite lamellae. Mechanical properties of the other cast iron types are somewhere between GJL and GJS [7]. The reasons why steel products continue being replaced with ductile cast iron castings include their good mechanical properties and lower economic demands of casting production.

Conditions favourable for the development of casting defects may arise in the casting process and solidification of cast iron castings. These defects have various impacts on the final cast piece quality.

Defects that impair the structure integrity, thereby worsening the characteristics of the final cast piece, are especially serious. As a rule, these defects cannot be removed and therefore they result in disposal of the given piece and the related financial losses. Porosity is one of the most common defects in castings and is caused by metal shrinkage in the solidification process. However, the task of minimising porosity is complex and requires consideration of chemical composition of the material, the casting temperature, the inoculation method, cooling capacity of the mould, risers, etc. [8, 9].

Addressing the causes of defects has been regarded as one of the most important aspects in foundries. Ideally, the causes are eliminated before initiating production. For example, numerical simulations can be used for this purpose. When the model is successfully validated, numerical simulation can be an effective tool for a relatively prompt prediction of changes in the casting structure after various technological parameters are modified. Current software systems offer good precision when predicting the course of filling [10, 11], solidification [12, 13], tension states [14-16] in the castings, but also other factors such as porosity [13, 17] and susceptibility to cracking and formation of fractures.

The purpose of the presented research was to optimise the production of the casting 902 319 GEHAUSE from cast iron with spheroidal graphite according to ČSN EN 1563. This casting is intended for manufacturers of ship manipulators where a superior level of material processing is expected. The used materials must comply with rigorous criteria for their quality and adherence to chemical and mechanical properties at low temperatures. A complete numerical analysis of the casting was performed in the MAGMASOFT® for this purpose; the analysis focused on filling, solidification and tension states in order to optimise the production technology and minimise any risks of porosity, cracks and fractures.

2. Material characteristics

The casting 902 319 GEHAUSE which was the subject of numerical simulations, represents components designed for ship manipulators for medium-sized ships. Therefore, a high level of material quality is required for this casting to achieve chemical composition, desired structure and mechanical properties. The gross weight of casting is 640 kg and wall thicknesses range between 24-48 mm.

The casting is made of the material EN-GJS-400-18-LT, i.e. cast iron with spheroidal graphite and a ferritic matrix with a guaranteed impact work value at -20°C [18, 19]. Chemical composition of the material EN-GJS-400-18-LT is provided in **table 1**.

The material EN-GJS-400-18-LT offers outstanding machinability, combined with resistance to low temperatures, good fatigue strength, electric conductivity and magnetic permeability, including enhanced ductility and an ability to absorb noise and vibrations well. Mechanical properties of the material EN-GJS-400-18-LT are shown in **table 2**.

Table 1. Chemical composition of the material EN-GJS-400-18-LT

	C	Si	Mn	P	S	Cu	Mg	Cr
Min.	3.50	1.90	0.10	×××	×××	×××	0.040	×××
Max.	3.60	2.00	0.15	0.040	0.010	0.05	0.060	0.02

Table 2. Mechanical properties of the material EN-GJS-400-18-LT (standard wall thickness $30 < t \leq 60$ mm) [18]

Tensile strength	Yield strength	Ductility	Brinell hardness	Min. absorbed energy (low temperature (-20 ± 2) °C)
Rm (MPa)	$Rp_{0.2}$ (MPa)	A_5 (%)	HB (-)	KV_2 (J)

Min. 380	Min. 230	Min. 15	Min. 130	12

3. Geometry and numerical simulation settings

Geometry of the casting 902 319 GEHAUSE, together with the inlet system and other components of the casting system (the inlet system, the core, risers, vents, etc.), is shown in **figure 1**. The mold is always formed for 1 piece of casting. A computational mesh was generated for the geometry with a constant size of 5 mm of the computational mesh element chosen for all elements of the casting system (except the mould). The resulting mesh for the casting 902 319 GEHAUSE, composed of 1,083,135 elements, is presented in **figure 1**.

The input parameters for numerical simulations of the casting were defined based on operating data supplied by KOVOSVIT MAS Foundry, a.s. These parameters include chemical composition of the material EN-GJS-400-18-LT (see **table 1**), the casting temperature, the inoculation method, materials to make the mould, the casting filling time, etc. The cast iron with spheroidal graphite used to make the given cast piece type is inoculated in the operating conditions using the pouring method in two phases: pre-inoculation in the ladle and block inoculation in the pouring basin. Single-phase inoculation in the ladle was considered in the calculation given the limits of the MAGMASOFT®.

Figure 1. Geometry and computational mesh of the casting 90 319 GEHAUSE and the inlet system

The calculation of filling, solidification and tension states was carried out in the simulation. Continuous filling of the mould was ensured using the Automatic Filling Control feature that regulated the cast iron flow by controlling the cast iron level in the pouring basin. To calculate the tension, the time of removing the casting from the mould (24 hours) and the time of core shake out from the casting (50 s from casting shake out from the mould) were also defined. A thus prepared simulation was run and successfully completed. The calculation time was about 3 days.

4. Results analysis

This section presents selected results of the filling, solidification and casting tension states simulation. Considering the amount of assessed variables, predominantly those casting areas are presented below which are associated with a risk and could potentially cause quality degradation.

The temperature field of the cast piece filling process is shown in **figure 2**. The molten metal temperature does not decrease below the liquidus temperature. Thus, cold connection defects can be excluded. The velocity field of the cast piece filling process is shown in **figure 3**. Higher flow velocities are achieved in the gates, resulting in metal splashing into the casting volume; this might lead to air being captured in certain parts of the mould (indicated in red).

Figure 2. Temperature field (°C)

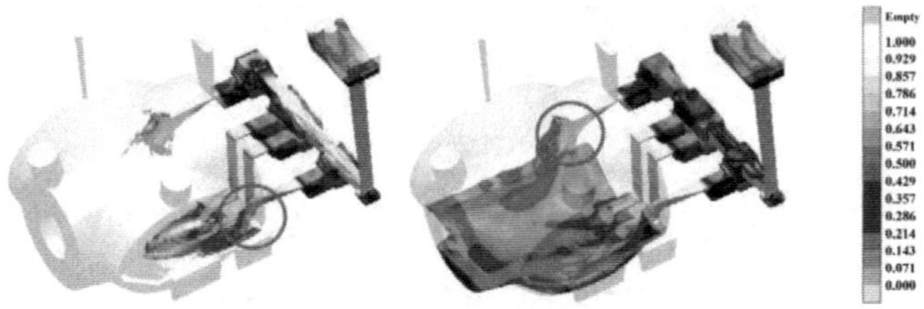

Figure 3. Velocity field (m.s^{-1})

The casting solidification process is illustrated in **figure 4** through the liquid phase loss. Isolated islets of molten mass develop during the solidification process. The solidification characteristics are not optimal particularly in the flange (indicated in red) above which only a vent is located. There is an actual risk of porosity development in this area even considering its suppression by graphite expansion. It would be advisable to place a feed instead of a vent in this area. Evidence of unsuitable solidification characteristics under the vents is also provided by the feeding module as illustrated in **figure 5**. As indicated by the feeding module, the molten metal under the vents solidifies more slowly compared to the vents. This results in development of the above mentioned undesirable molten metal islets. In particular, the area indicated in red in **figure 5** is associated with a risk, where porosity could develop. The situation under the second vent is similar, although less critical. Porosity found in this area should probably be suppressed through graphite expansion.

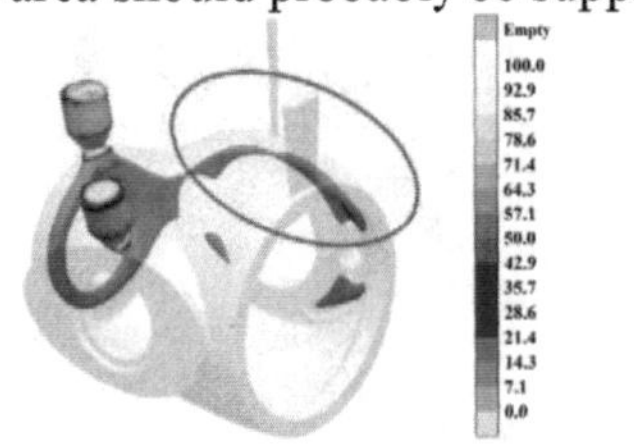

Figure 4. Solidification characteristics of the casting – liquid fraction (%)

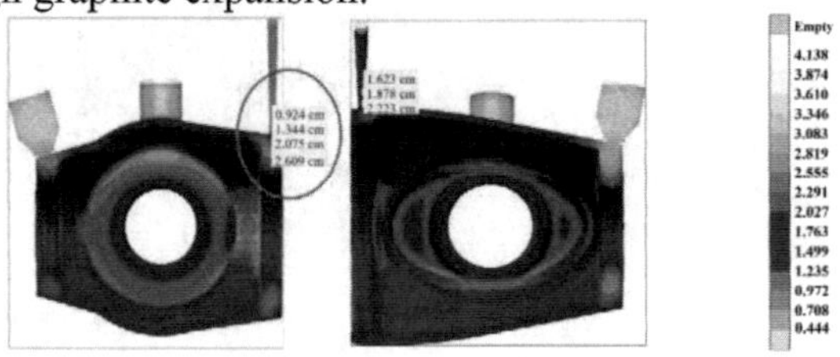

Figure 5. Feeding module (cm)

Temperature nodes develop in the solidification process as illustrated in **figure 6**. Some of the temperature nodes reach the casting surface (indicated in red). There is a risk of casting surface cave-in at these points. Porosity at the end of the solidification process is shown in **figure 7**. The results indicate points where more than 10% of the metal is missing. Porosity may form at these points.

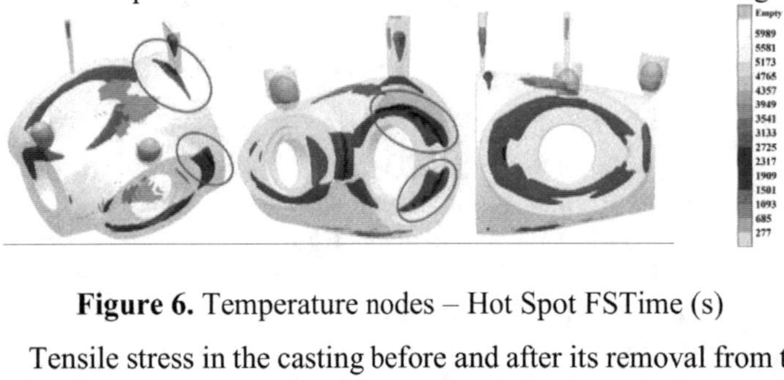

Figure 6. Temperature nodes – Hot Spot FSTime (s)

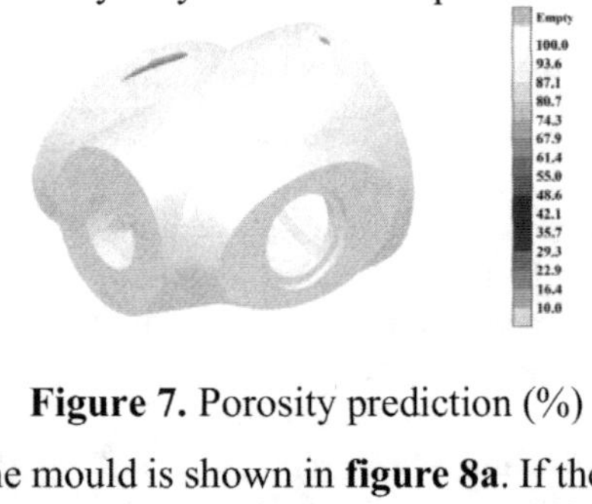

Figure 7. Porosity prediction (%)

Tensile stress in the casting before and after its removal from the mould is shown in **figure 8a**. If the casting stays in the mould, a relatively high tensile stress is accumulated, which may exceed the tensile strength of the material. However, the tensile stress is dispersed by other types of stress the casting is subjected to, and thus the situation need not be as serious in the end. Nevertheless, this state is not beneficial for the casting, either, as the higher residual tension increases the risk of

IOP Conf. Series: Materials Science and Engineering 1243 (2022) 012006 doi:10.1088/1757-899X/1243/1/012006

cracking should the tension exceed the tensile strength of the material. After the casting is removed from the mould, the tension is relieved as illustrated in **figure 8b**, and the tension decreases even further when the casting cools down to the ambient temperature. Therefore it can be recommended to reduce the casting retention time in the mould to a duration that is absolutely necessary. Based on comparison of the maximum tensile stress in the casting and the tensile strength of the material, it can be said that residual tensions present in the casting after its shake out do not exceed the minimum tensile strength and thus there is no risk of material impairment.

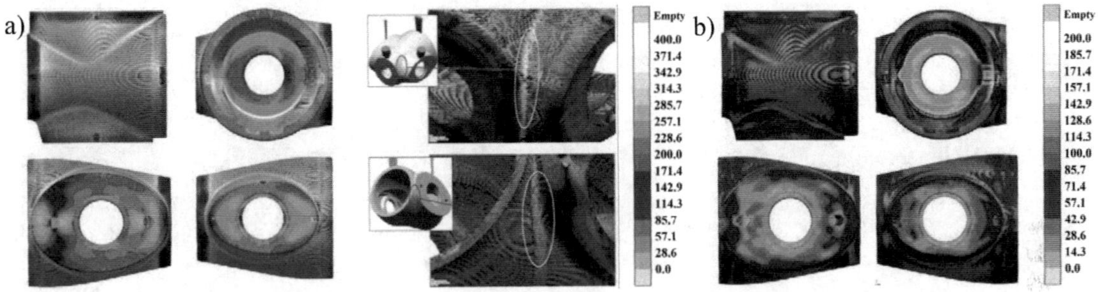

Figure 8. Tensile stress (MPa) – a) before shake out, b) after shake out

The susceptibility to cold cracking was assessed using the Cold Crack criterion. As illustrated in **figure 9**, there is no point in the casting where the critical value 1 is markedly exceeded. Thus the casting should not crack. However, the radius area on the inner side of the casting seems to be a critical point. High Cold Crack criterion values are found at this place. The casting should not crack under the current conditions. However, if some of the process parameters are unsuitably changed, cracks might form precisely at this point. The risk can be suppressed by increasing the radius at this point if allowed by the structural design of the casting.

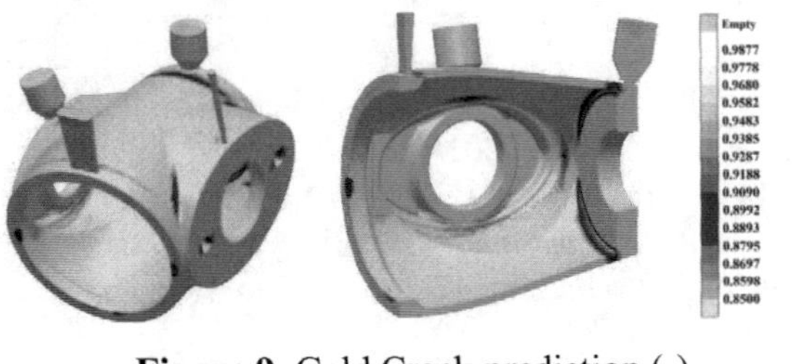

Figure 9. Cold Crack prediction (-)

5. Conclusions

Numerical analysis of casting the 902 319 GEHAUSE cast piece type was carried out based on operating data gathered in cooperation with technologists of KOVOSVIT MAS Foundry a.s. The numerical simulation did not reveal any serious defects of the casting production technology. Based on the numerical simulation, the conclusions can be summarized as follows:

- The molten metal temperature does not decrease below the liquidus temperature in the mould filling process. The set casting temperature is optimal and there is no risk of defects such as cold shut defects and insufficient filling.
- Mould filling is correct in terms of the casting velocity. However, higher velocities are reached in the gates, resulting in metal splashing into the casting volume. In order to optimise the filling

process, it can be recommended to adjust the cut shapes in order to prevent metal splashing and to minimise the risk of air being captured in certain parts of the mould.

- The solidification process of the casting is not optimal. Porosity development can be expected in flanges under the vents. It is therefore recommended to consider replacing the vents with adequate risers. Given the current technological conditions, there is a risk of porosity development in several areas of the casting volume.
- Temperature nodes develop in the casting, and they reach the surface at some points. There is a risk of casting surface cave-in at these points.
- The casting tension before and after its removal from the mould should not exceed the tensile strength of the material. Despite that, it is recommended to reduce the casting retention time in the mould during which the casting is exposed to tensile stress (at some places) that approaches the tensile strength of the material.
- No cracking will occur in the casting. However, a place potentially associated with a risk of cracking was identified. This risk can be suppressed through a slight modification of the casting geometry in accordance with its structural design.

Based on the numerical simulation results, we shall further focus on an optimisation of the casting technology in the operating conditions in order to reduce the risk of defects in 902 319 GEHAUSE casting types.

Acknowledgements
The paper was prepared under the support of the Czech Ministry of Industry and Trade within the scope of the TRIO programme, as part of project Reg. No. FV40034 "Research and development of complex technology of castings production from high-quality ductile irons castings". The authors thank the MAGMA team for preparing master studies in MAGMASOFT® for the above project.

References
[1] Census of World Casting Production: Total Casting Tones Dip in 2019. *Mod. Cast.* 111(1) pp 28-30
[2] Soiński M S, Kordas P and Skurka K 2016 Trends in the Production of Castings in the World and in Poland in the XXI Century *Arch. of Foundry Eng.* 16(2) pp 5-10.
[3] Menk W, Tunzini S, Rieck T, Honsel Ch and Weiss K 2010 Development of Ductile Iron Material, Simulation and Production Technology to Locally Strengthen Castings. *Key Eng. Mater.* 457 pp 343-348
[4] Soiński M S, Kordas P and Skurka K 2016 Trends in the Production of Castings in the World and in Poland in the XXI Century *Arch. of Foundry Eng.* 16(2) pp 5-10.
[5] Laine J, Jalava K, Vaara J, Soivio K, Frondelius T and Orkas J 2021 The Mechanical Properties of Ductile Iron at Intermediate Temperatures: The Effect of Silicon Content and Pearlite Fraction *Int. J. Metalcast.* 15(2) pp 538-547
[6] ČSN EN 1560. Foundry – Designation system for cast iron – Material symbols and material numbers (Praha: The Czech Office for Standards, Metrology and Testing) p. 16.
[7] Pero-Sanz Elorz J A, González D F and Verdeja L F 2018 Physical Metallurgy of Cast Irons
[8] Alagarsamy A and Kumar S 2020 Demystifying Shrinkage Porosity in Ductile Iron Castings *Int. J. Metalcast.* 14(4) pp 1136-1151
[9] Regordosa A, Llorca-Isern N, Sertucha J and Lacaze J 2018 Evolution of Shrinkage with Carbon Equivalent and Inoculation in Ductile Cast Irons *Mater. Sci. Forum.* 925 pp 28-35
[10] Hawranek R, Lelito J, Suchy J S and Zak P 2009 The Simulation of a Liquid Cast Iron Flow through the Gating System with Filter *Arch. Metal. Mater.* 54(2) pp 351-358.
[11] Patil R T, Metri V S and Tambore S S 2015 Analysis and Simulation of Die Filling in Gravity

IOP Conf. Series: Materials Science and Engineering 1243 (2022) 012006 doi:10.1088/1757-899X/1243/1/012006

Die Casting using MAGMA Software *Int. J. Eng. Res. Technol*. 4(11) pp. 556-559
[12] Bhatt H, Barot R, Bhatt K, Beravala H A and Shah J 2014 Design Optimization of Feeding System and Solidification Simulation for Cast Iron *Proc. Tech.* 14 pp 357-364
[13] Khan M A A and Sheikh A K 2018 Mechanical characterization and quality of iron castings using optimized mold design: simulations and experimental validation *Int. J. Adv. Manuf. Tech*. 98(1-4) pp 799-809
[14] Sonne M R, Frandsen J O and Hattel J H 2015 Comparison of residual stresses in sand- and chill casting of ductile cast iron wind turbine main shafts *IOP C. Ser-Mat. Sci* 84
[15] Ignaszak Z 2010 Towards optimization of stress simulation in real casting-mould system *Arch. of Foundry Eng.* 10(4) pp 69-76
[16] Smith F, Markgraaff J, Marais D and Venter A M 2018 Residual Stress Determination of Ductile Cast Iron by means of Neutron Diffraction *Mater. Res. Proc.* 4 pp 79-84.
[17] Singh S, Sharma N and Khanna R 2020 Defect elimination in green sand casting using simulation techniques: A case study *Int. J. Model. Simul. Sci. Comput*. 11(5)
[18] ČSN EN 1563. Founding – Spheroidal graphite cast irons (Prague: The Czech Office for Standards, Metrology and Testing) p. 40.
[19] ČSN EN ISO 945-1. Microstructure of cast irons - Part 1: Graphite classification by visual analysis (Prague: The Czech Office for Standards, Metrology and Testing) p 68.

IOP Conf. Series: Materials Science and Engineering 1243 (2022) 012007 doi:10.1088/1757-899X/1243/1/012007

Options for Implementing Additive Manufacturing Technologies into a Foundry for Small Castings

F Votava[1*], D Bricín[1]

[1]University of West Bohemia, Regional Technological Institute, Univerzitni 8, 301 00, Pilsen, Czech Republic

[*]Email: votavaf@rti.zcu.cz

Abstract. This paper is focused on the possibility of the implementing various additive manufacturing technologies into the investment casting process. The most frequently used AM technologies were employed to print fusible models of valves for musical instruments. This was done to accelerate the production process of new products in the foundry. The output of the work should determine the optimal AM technology according to the quality of surface and shape for creating a fusible model and suitable materials should also be recommended. Furthermore, the financial aspects of the variants were evaluated. For this evaluation, the analyses of the ash content, visual inspection, surface roughness and topography of the printed models were done. After these analyses, the models were formed, moulds were fired and the tin alloy was cast. Castings were subjected to the same analyses as for the models. X-ray inspection was done and finally the defects of the castings were evaluated.

1 Introduction

Investment casting is used to make complex castings with good surface quality and small dimensional tolerances. It is one of the oldest known casting techniques. In many cases, castings do not even have to go through finishing operations because their dimensional accuracy is sufficient for their final use [1, 2, 3]. Despite these advantages, it is a time-consuming, laborious process, but it can create complex components with high added value for specialized industries in a wide range of alloys [4].

The reason for this work was to find a new way of production in a foundry for small castings. This foundry is a part of a company that specializes in the production of musical instruments. Castings are produced by the investment casting method. This foundry produces mainly castings of small mechanical valves for wind instruments, e.g. saxophones, clarinets, etc.

The whole complex process of investment casting was shown to us in the foundry. From the production of the parent model, through the production of the silicone mould for wax models, creation of the mould, its firing and then final casting. According to the knowledge acquired from our research possible ways of implementing AM (additive manufacturing) technologies will be designed. Production of one type of parent valve (for the creation of a silicone mould) can take up to 8 hours, depending on the complexity of its shape. Moreover, different techniques are needed to create the final shape, as shown in figure 1 – three parts have to be joined together, (1) cast, (2) milled and (3) turned part.

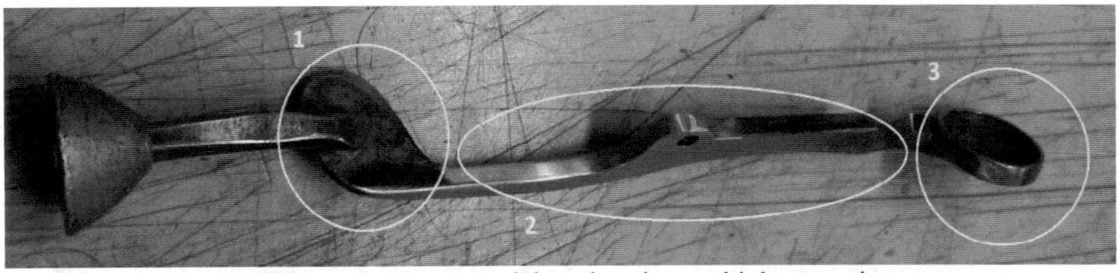

Figure 1. Parent model made using multiple operations.

Two different types of valves (shown in figure 2) further served as an input for the design of virtual models that will be used for AM technologies.

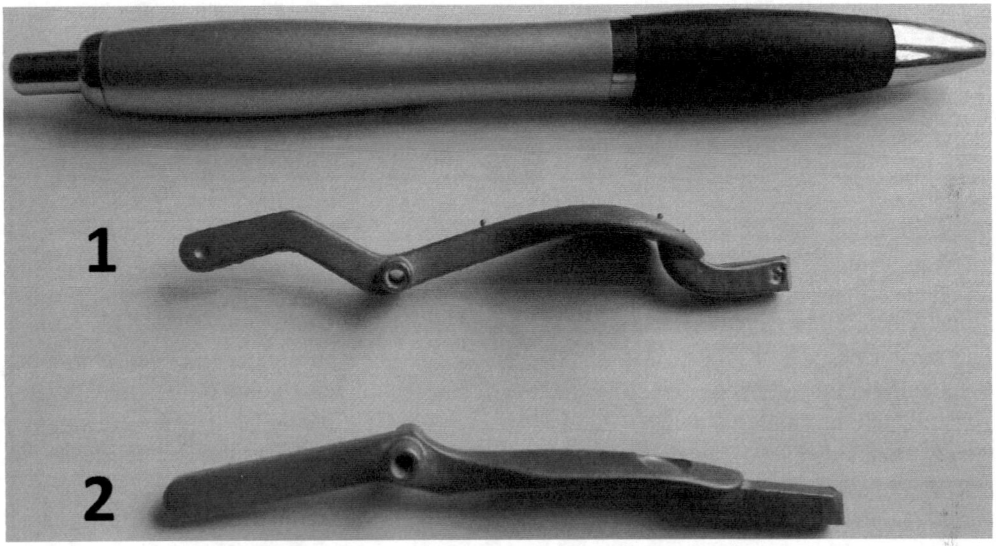

Figure 2. Two different models of bronze valves.

2 Options for implementation

Cast models were converted into digital form (STL files) in two different ways. The first consisted of measuring the model with a sliding gauge and then creating the virtual model in a CAD (Computer-Aided Design) program. The second consisted of using a 3D measuring arm with a scanner. The second approach was chosen because of the more complicated shape of the valve and to demonstrate another possibility of conversion to a virtual model. Figure 3 shows both STL models which are further used for 3D printing.

IOP Conf. Series: Materials Science and Engineering 1243 (2022) 012007 doi:10.1088/1757-899X/1243/1/012007

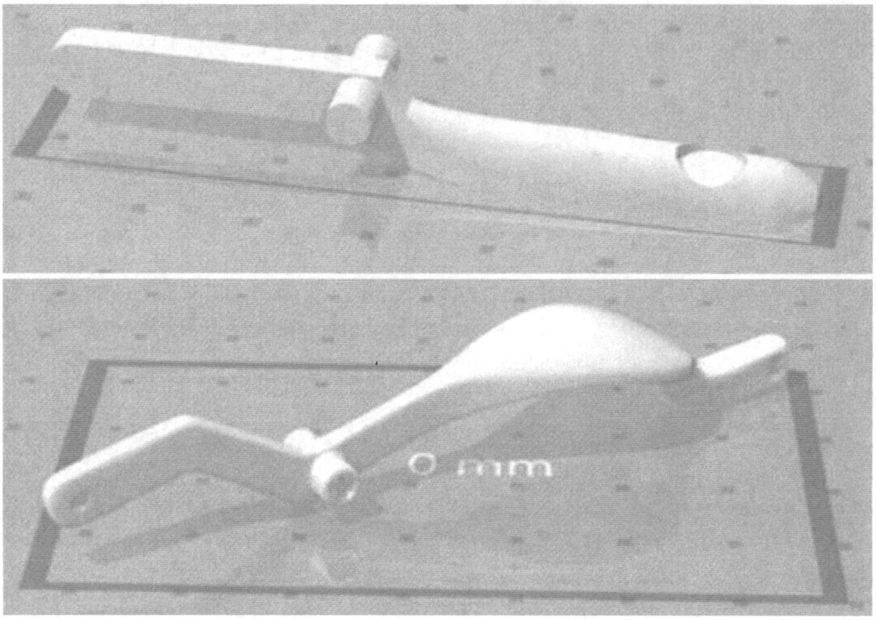

Figure 3. STL models – upper valve created in CAD program, lower valve scanned by a 3D measuring arm.

Below are shown many options for implementing additive technologies into the process of investment casting. In the case of small production, this action can reduce costs associated with production. In the case of custom production, it can reduce costs and significantly shorten the time of the whole production process, as shown in figure 4. The difference in time consumption of the whole process between 'conventional investment casting' and approaches that use rapid prototyping techniques is huge [5].

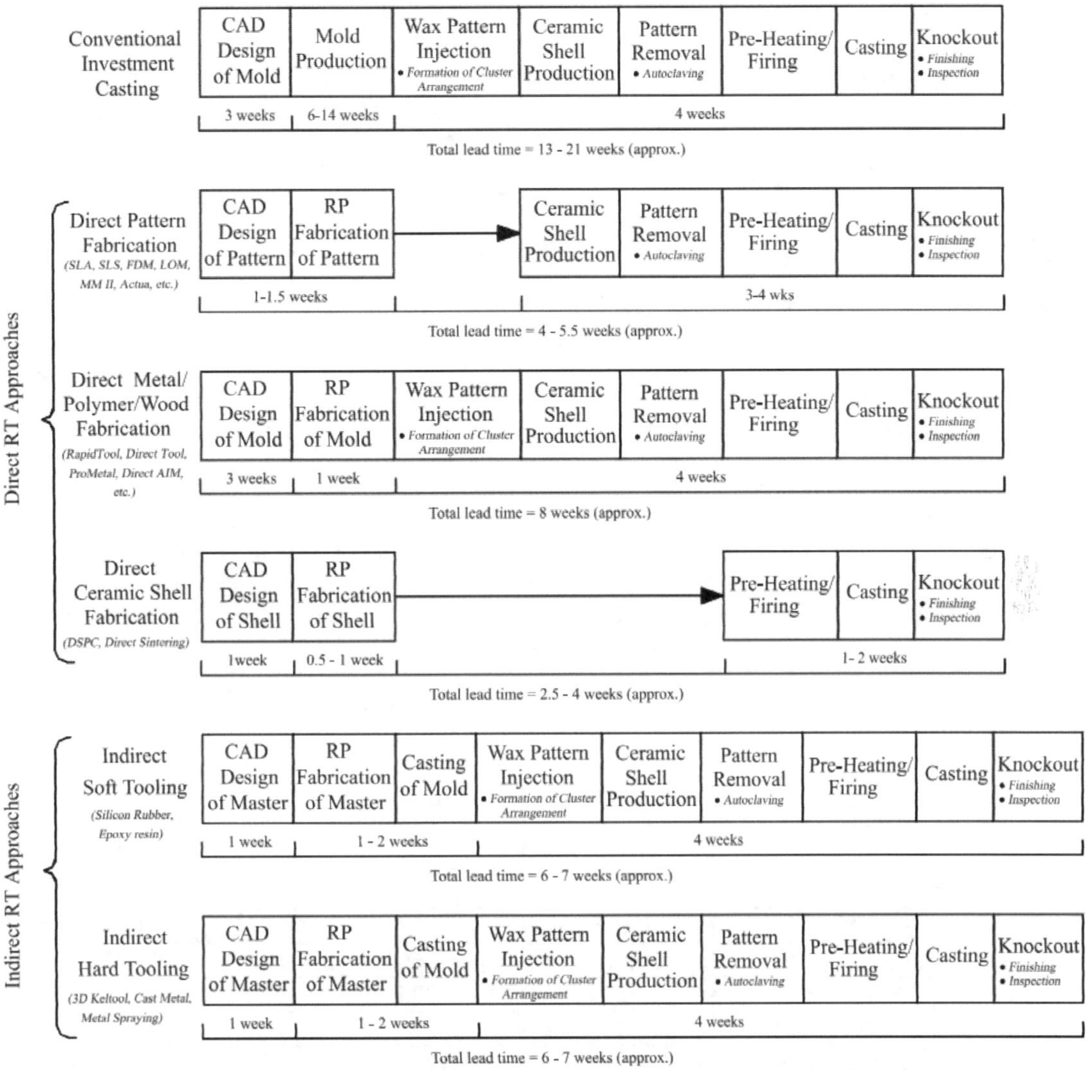

Figure 4. Comparison of conventional investment casting (IC) and IC with the implementation of AM technologies [5].

According to figure 4, there are two approaches for the implementation of AM technologies into the process – direct and indirect. This work focuses only on the 'direct pattern fabrication' approach using different AM technologies: stereolithography (SLA), fused deposition modelling (FDM), multi-jet modelling (MJM) and multi-jet fusion (MJF). These technologies were chosen because FDM and SLA are the most widespread printers for polymers and wax. On the other hand, MJM and MJF technologies were also chosen because of their availability. Then it would be possible to compare commonly used printers with professional ones. The models were printed on these devices and different materials were used. The most often printed materials (e.g. PLA – Polylactic acid, ABS – Acrylonitrile butadiene styrene) were used for each printing technology, and two special materials were purchased for subsequent casting (Castable Wax and MoldLay). The reason for this was to find the best printer and material for this direct approach [5, 6].

With different approaches, it would be possible to print e.g. parent model, a mould for creating wax models, a mould for casting (also called a ceramic shell), etc.

3 Experimental part

3.1 Printers and materials

Models were printed using 4 different methods – SLA, FDM, MJM and MJF. Six different devices were used (3 different printers for the FDM method – to compare the quality of this most widespread method and different materials). The printers were: for FDM: Geetech A10M (materials: PLA, MoldLay); Original Prusa i3 MK3S (PLA, ABS); Ultimaker S5 (PVA – Polyvinyl alcohol); for SLA: Original Prusa SL1 (Castable Wax); for MJM: ProJet 3510HD (M3 Crystal); for MJF: HP Jet Fusion 4200 (PA 12 – Polyamide 12). Printing settings differed. The main goal was to print as finely as possible. All printing temperatures were determined according to the material list provided by the producer. Only in the case of MJF is the input material powder, MJM and SLA use resin material, and all FDM printings use filament.

After all the models have been printed, further analyses follow, such as ash content analysis, surface roughness and topography, visual inspection, and removal of supports.

3.2 Ash content analysis

Ash content analysis was done to find the proportion of non-combustible components of each material. It was done according to a firing curve used in the foundry. The maximum temperature reached was 720 °C for 30 minutes. For this analysis we used a precision laboratory scale Radwag AS 220/C/2, glazed ceramic cups and an atmospheric chamber furnace with a heating rate of 10 °C/min. The results of this analysis are shown in table 1.

Table 1. Ash content analysis.

Material	1. measurement	2. measurement	Arithmetic mean	Standard deviation [±]
	Ash content [%]			
Flexible Blue	0.046	0.028	0.037	0.009
Castable Wax (SLA)	0.095	0.066	0.081	0.015
PLA (FDM)	0.395	0.385	0.39	0.005
ABS (FDM)	0.263	0.282	0.273	0.009
MoldLay (FDM)	0.35	0.365	0.358	0.008
M3 Crystal (MJM)	0.023	0.032	0.028	0.005
PA 12 (MJF)	0.442	0.439	0.441	0.002
PVA (FDM)	0.067	0.06	0.064	0.004

The first material (Flexible Blue) in table 1 served as the reference material because it is used in the foundry for the production of wax models. Only M3 Crystal (MJM) achieved lower ash content so it seems to be an ideal material for use in the foundry. It is possible to say that Castable Wax (SLA) and PVA (FDM) could also be suitable because the ash content did not exceed 0.1 %. Other materials have a higher content of ash and they seem to be less suitable for using in the foundry. It is worth mentioning ABS (FDM) because it was the most difficult material to remove from the cup. This phenomenon could result in the bad surface quality of the final casting because the molten metal would not absorb the ash inside, creating surface defects. Even more, ABS does not seem to be suitable because carcinogenic styrene is released into the atmosphere when it is fired [7].

Another important aspect was infill. In the case of SLA, MJM and MJF materials it was not possible to print internal structures with cavities, so models were 100 % filled. In the case of FDM printings, two types of internal structures were used. Both structures were made according to a reference from a study focused on the collapsibility of 3D printed parts. The first inner structure consisted of large

inner-connected quadrilateral cells with constant wall thickness [8]. As stated in another study, a second inner structure could be even more suitable for this use – octagonal cells [9]. The reason for creating inner hollow structures is simple – during the firing in the furnace, the ceramic shell can crack. This can happen due to different thermal expansions of the material used for the model and the material of the ceramic shell. When the inner hollow structure is created during the firing it starts to collapse into itself before it can destroy the ceramic shell [10].
The actual ash content may be lower after firing than this analysis indicates because during firing the material can flow out of the mould cavity.

3.3 Surface roughness and topography

Visual inspection and measurement of roughness were done next. Roughness was analysed on a Hommel-Etamic T8000 using a 2 μm diamond tip with an apex angle of 90°. Four measurements were made for every printed model: two for length and two for the width of each model. Results from the roughness tester were also supported by surface topography which was measured on an optical-scanning microscope (IFM G4) with a 10x objective, a vertical resolution of 2.0249 μm and a lateral resolution of 1.9012 μm. Figure 5 shows a comparison of surface topography between FDM and SLA printed valves. At first sight, there is a big difference between these models. The height subrange is displayed in colour from - 100 to 100 μm. In the first sample, almost the entire height subrange is covered (appearance of all colours). In the second case, the values go approximately from - 30 to 30 μm (almost uniform surface).

Figure 5. Surface topography of printed models (upper – FDM; lower – SLA).

Analyses showed that the most suitable printed material should be Castable Wax (SLA), then M3 Crystal (MJM) and PA 12 (MJF). All FDM models were coarser because this method does not have such a precise resolution of one layer and it uses supports to print the models. After the removal of the supports, it can still be seen that the surface is not smooth as shown in figure 6. Figure 6 shows the difference between the red valves made of PLA printed on Geetech A10M (FDM) and the purple valves made of Castable Wax (SLA).

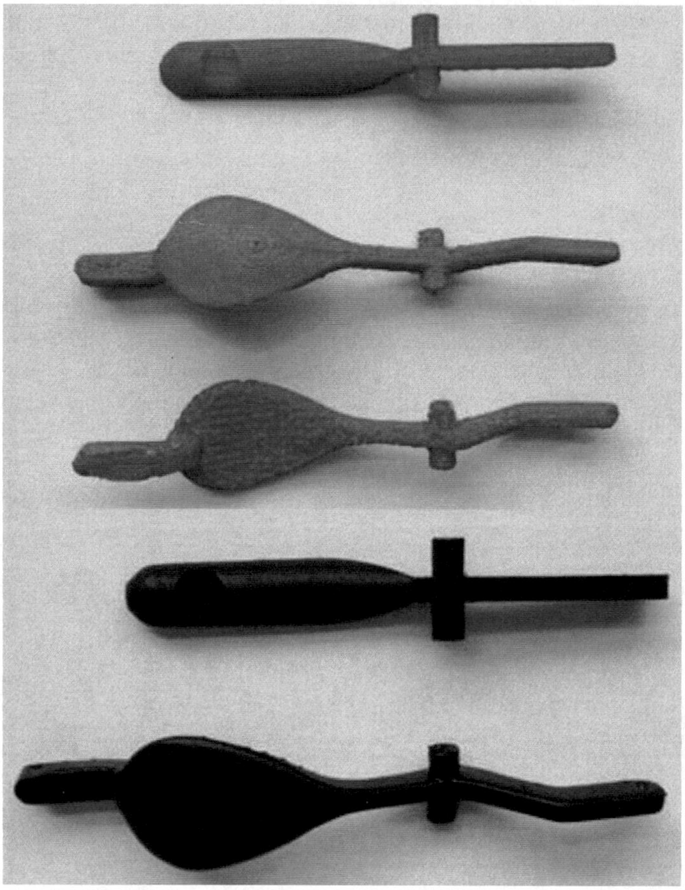

Figure 6. Difference between FDM (red) and SLA (purple) printed valves.

3.4 Ceramic shell production, firing and casting

Each printed model was glued to an inlet system, shown in figure 7. One universal inlet system was designed and it was printed from ABS (FDM) with a layer thickness of 0.4 mm to make printing as fast as possible. In the foundry, several models are always connected to a large inlet system. For the experiment, it was decided to prepare a mould for each model because while the mould is fired, it may crack. The mould was made of material directly intended for this purpose – "Gold Star XL" plaster. This gypsum mixture was weighed and mixed with water. The ratio of water to gypsum was 4:10. Then the mixture was poured into the prepared forms and each model with an inlet system was inserted inside one of those forms. All moulds were tapped to reduce bubbles inside the mixture. It started to solidify quickly. After two hours the solidified gypsum was removed from the forms, and the inlet systems were removed (to not affect the different materials used for models), and the gypsum moulds were marked and put into the chamber furnace. The firing was done according to the curve shown in figure 8.

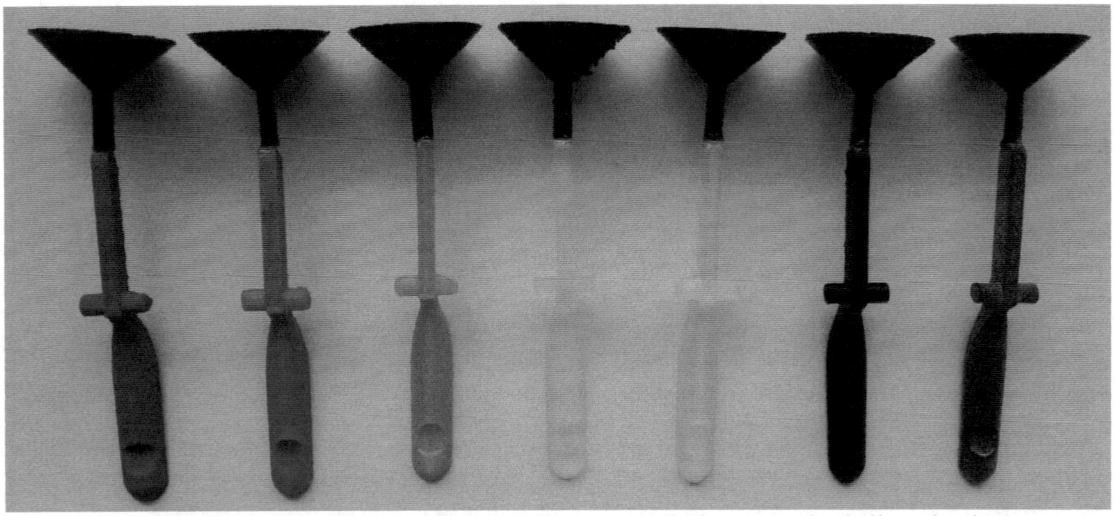

Figure 7. Models attached to the inlet system (before ceramic shell production).

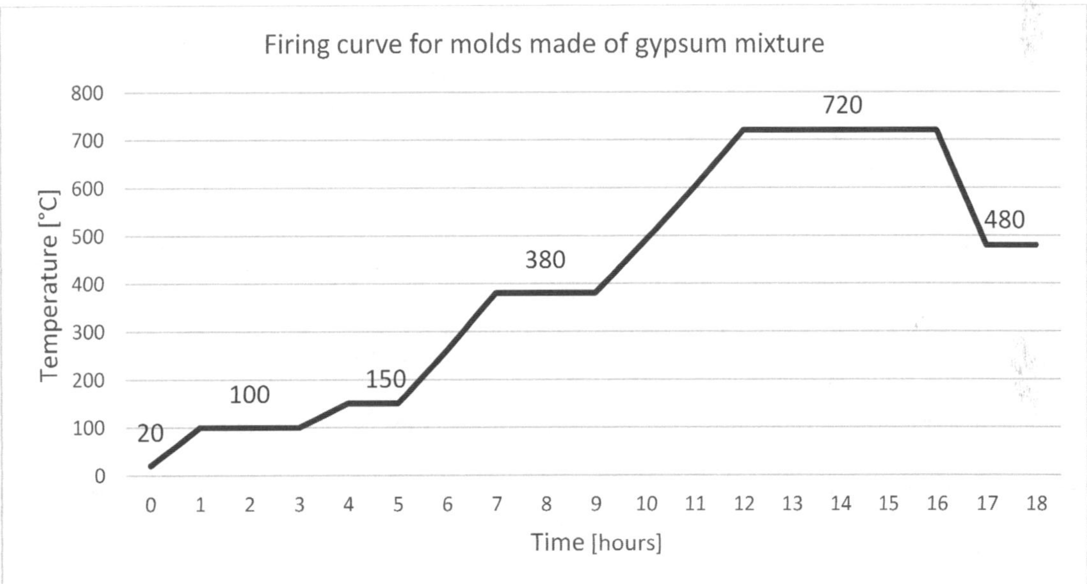

Figure 8. Firing curve for moulds made of gypsum mixture (Temperature/Time).

After removing the moulds from the furnace, tin alloy was poured inside when the temperature of the moulds was around 300 °C. Tin alloy was chosen because of the low melting point of tin (232 °C). Two series of castings were made. In both series, one mould out of seven cracked. Both cracked moulds contained a 100 % infilled model (the first one from MJM and the second one from MJF). Figure 9 shows all the cast models with the inlet system (from the first casting). The black circled parts of figure 9 are grooves that are there because of their insertion into the gypsum mixture that was solidifying quickly. Based on this, the process of preparing the moulds for the second casting was modified. The models were not inserted into the mixture, but each model was poured with the mixture inside the form to avoid the formation of grooves.

IOP Conf. Series: Materials Science and Engineering 1243 (2022) 012007 doi:10.1088/1757-899X/1243/1/012007

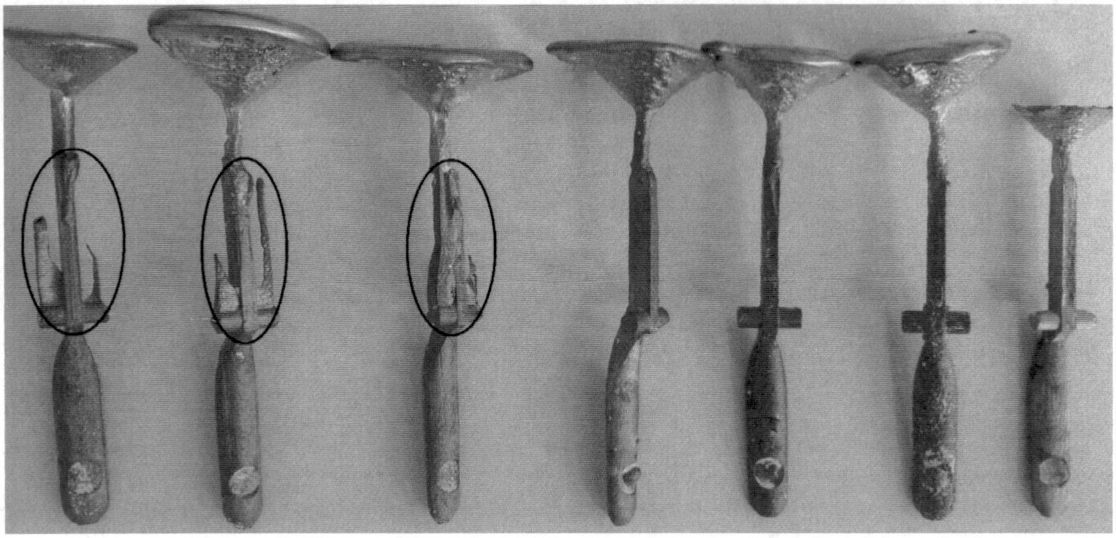

Figure 9. Cast models with the inlet system made of tin alloy (1st casting).

3.5 Defects of castings

After the models were cast, visual inspection, surface roughness and topography were done. Also, all the castings were subjected to non-destructive X-ray inspection. Only a very few internal defects with very small dimensions were found and they would not affect the use of the components. The most important step was the evaluation of the defects according to ČSN 42 1240 [11].

The evaluation was done for every printed valve. Figure 10 shows the most common defects that appeared on almost all the castings. Further, in figure 10 can be seen layers from 3D printing because these valves were printed using FDM. Models made by SLA, MJM and MJF had smoother surfaces according to the results from surface roughness measurement.

Shots appeared on the castings because when the model was formed, bubbles of air were trapped on the surface of the model even if the form was tapped. The defect called "cake" is a consequence of the residual ash content inside the cavity of the ceramic shell that could stay stuck to the surface. Metal-free spaces are connected with the bad placement of the model in the mould and because of gravity casting conditions. The inlet system probably solidified earlier than the model. Otherwise, there would be enough molten metal in the inlet system that would add the metal to shrinking areas and metal-free spaces would not appear. Other defects were connected with broken moulds.

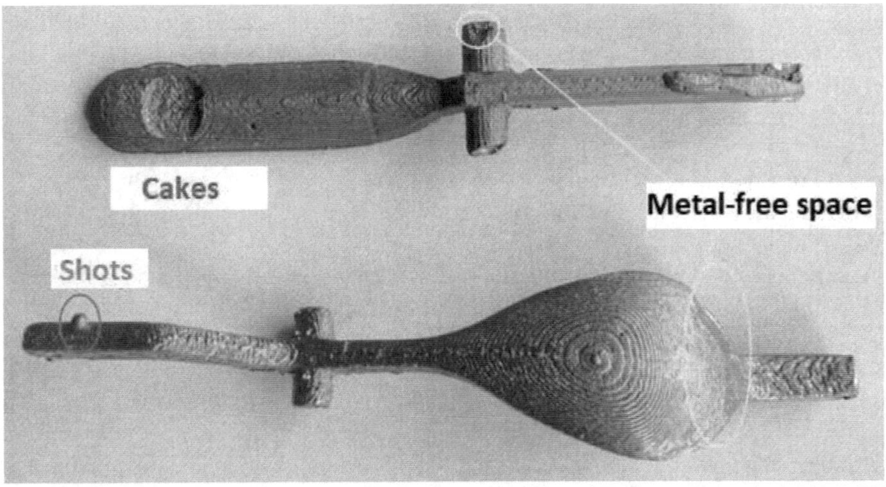

Figure 10. Examples of defects on valves from models printed on Geetech A10M (FDM) made of PLA.

4 Economic aspect

This part focuses on the economic aspect of the variants. It was simplified to operating costs in the form of the material used and the acquisition costs for the equipment, shown in table 2. Other costs, especially maintenance costs, energy, production area, etc. will be completely different for each printer.

Table 2. Acquisition costs of printers.

Equipment	Price incl. VAT [USD]
Geetech A10M (FDM)	345
Original Prusa i3 MK3S (FDM)	784
Ultimaker S5 (FDM)	7 355
Original Prusa SL1 (SLA)	1 725
ProJet 3510 HD (MJM)	77 900
HP Jet Fusion 4200 (MJF)	200 000

FDM printers are the cheapest option, SLA is more expensive, but acquisition costs for both methods reach a maximum of thousands of dollars. MJM and MJF methods are professional printers and the acquisition costs are in the region of hundreds of thousands of dollars and also materials are much more expensive.

5 Results and discussion

From the point of view of materials with the lowest ash content, the optimal variants are M3 Crystal (MJM), PVA (FDM) or Castable Wax (SLA). But in comparison with the material used in the foundry, only M3 Crystal has a lower ash content. The other materials have a higher proportion of non-combustible components (up to 11 higher than Flexible Blue – for PA 12).

During the visual inspection, the smoothest surfaces were from SLA, MJM and MJF. This fact was also confirmed when the roughness was measured. In comparison to conventionally cast valves, all AM technologies had coarser surfaces. FDM illustrates the problem with roughness depending on layer thickness. Also, the limiting aspect was that three methods (SLA, MJM and MJF) used 100 % infill. In the MJM and MJF models, the moulds cracked during the firing. Castable Wax (SLA) is designed for this purpose, so there was not that problem.

In terms of the visual inspection of castings, the same defects appeared in all the samples. This was caused by the conditions for gravity casting and also the bad design of the inlet system. This experiment was done unfortunately only with simple tools, not with the specialized tools and machinery used in a foundry. It can be assumed that if the casting was done in the foundry, the quality of the castings would be much higher (fewer defects, a better quality of surface, easier conditions for the moulding process, etc.) because, for example, the foundry is equipped with a vacuum device with a vibrating plate for the production of ceramic shells to reduce bubbles in the gypsum mixture and an induction vacuum furnace for casting.

From an economic point of view, the most suitable variants are: Geetech A10M and Original Prusa i3 MK3 (both FDM using PLA), then Original Prusa SL1 (SLA) using Castable Wax.

Table 3 was prepared based on the experiments and research described above. It includes all the parameters that influence the final casting and the costs associated with each type of production. The values were assigned subjectively to each parameter: the best (1), the worst (5). These values are based on the comparison of the results of analyses and experiments, which are then totalled. The lowest sum is achieved by the SLA method which means that it is the most suitable variant.

Table 3. Decision table.

Printer	Material	Ash content	Adhesion of ash	Quality of model	Surface of model	Quality of casting	Surface of casting	Acquisition cost	Material cost	Sum
Geetech A10M (FDM)	PLA	4	2	4	4	4	4	1	1	**24**
	MoldLay	4	2	4	4	4	4	1	3	**26**
Original Prusa i3 MK3S (FDM)	PLA	4	2	2	3	3	3	1	1	**19**
	ABS	3	5	5	5	5	5	1	1	**30**
Ultimaker S5 (FDM)	PVA	1	5	4	4	5	5	2	2	**28**
Original Prusa SL1 (SLA)	Castable Wax	1	2	1	1	2	2	2	5	**16**
ProJet 3510 HD (MJM)	M3 Crystal	1	2	1	1	3	2	4	5	**19**
HP Jet Fusion 4200 (MJF)	PA 12	5	5	2	2	3	2	5	5	**29**

6 Conclusion

The aim of the research was to find the optimal material and method of additive manufacturing technology for implementation into the process of investment casting. The direct method of printing the model was chosen. The experiment was done using four different printing technologies, on six AM printers, using seven different materials. The implementation should bring a reduction of costs and significantly shorten the time from start to final product. All analyses, measurements and economic aspects were accounted for in the decision table.

In summary, the optimal variant for the implementation of AM technology into the foundry for small castings is the Original Prusa SL1 (SLA) printer using Castable Wax. SLA achieves the best results if the intersection of all the outputs is considered. We believe that additional development of this implementation would lead to further improvements.

Certainly, the recommended implementation of additive technologies into the foundry is suitable mainly in the case of custom production (meaning different types of valves). The implementation starts losing meaning when the production switches to small series production mainly because of the longer time taken for the production of wax models compared to existing (conventional) production. On the other hand, AM technologies can print outside employee working hours. It would be possible to increase maximum production capacity using more printers (mainly if cheaper equipment is acquired).

The implementation of AM technologies into the process will remove the need for the production of a parent model and the production of a mould for wax models. Removing these two steps from the process can save labour and replace the expensive machines needed for these tasks (lathes, milling machines, vulcanizers, injection moulding machines, etc.). But at the same time, it can be stated that at the moment it will be more laborious to prepare the final surface of the castings. Therefore, further research should be carried out so that implementing AM technologies would fully replace the current steps of the process.

Acknowledgement
The present contribution has been prepared with the support of the student grant competition of University of West Bohemia in Pilsen, SGS-2021-025 Application of electron beam melting for welding of advanced high strength steels. The project was funded from specific resources of the state budget for research and devel-opment.

References

[1] Beeley, P R and Smart, R F 1995 *Investment Casting*. London: The Institute of Materials. ISBN 0901716669 pp 13-4
[2] Horáček, M 2009 *Rozměrová přesnost odlitků vyráběných metodou vytavitelného modelu*, Brno. [Online]. Available from: http://ust.fme.vutbr.cz/slevarenstvi/sites/default/files/clanky/technologie-vytavitelneho-modelu/technologie-vytavitelneho-modelu.pdf, [Accessed: 10. 9. 2021] p 3
[3] Herman, A 2007 *Lití na vytavitelný model*, Ústav strojírenské technologie, ČVUT v Praze. [Online]. Available from: http://u12133.fs.cvut.cz/assets/subject/files/32/presne-liti-na-vytavitelny-model.pdf, [Accessed: 10. 9. 2021] pp. 1-2
[4] Jones, S and Yuan, Ch 2003 *Advances in shell moulding for investment casting*. Journal of Materials Processing Technology. ISSN 0924-0136
[5] Cheah, C M, Chua, C K, Lee, C W, Feng, C and Totong, K 2005 *Rapid prototyping and tooling techniques: A review of applications for rapid investment casting*. Int. J. Adv. Manuf. Technol. 25. 308-320. 10.1007/s00170-003-1840-6 pp 308-20
[6] Lee, C W, Chua C K, Cheah, C M, Tan, L H and Feng, C 2004 *Rapid investment casting: Direct and indirect approaches via fused deposition modelling*. International Journal of Advanced Manufacturing Technology
[7] Azimi, P, Zhao, D, Pouzet, C, Crain, N E and Stephens, B 2016 *Emissions of Ultrafine Particles and Volatile Organic Compounds from Commercially Available Desktop Three-Dimensional Printers with Multiple Filaments; Environmental Science & Technology 50* (3), 1260-1268; DOI: 10.1021/acs.est.5b04983
[8] Harun, W S W, Sharif, S, Idris, M H and Kadirgama, K 2009 *Characteristic studies of collapsibility of ABS patterns produced from FDM for investment casting.* Materials Research Innovations
[9] Rahmati, S & Norouzi, Y and Hojjat, Y 2009 *A novel lattice structure for SL investment casting patterns*. Rapid Prototyping Journal
[10] Pattnaik, S, Jha, P K, and Karunakar, D B 2014 *A review of rapid prototyping integrated investment casting processes*. Proceedings of the Institution of Mechanical Engineers, Part L: Journal of Materials: Design and Applications
[11] ČSN 42 1240 1965 *Vady odlitků. Názvosloví a třídění vad*. Czech office for standards, metrology and testing, 44 p. Sorting character: 421240

IOP Conf. Series: Materials Science and Engineering 1243 (2022) 012008 doi:10.1088/1757-899X/1243/1/012008

Prediction of defects of ductile iron casting by numerical simulation

L Socha[1], K Gryc[1], J Sviželová[1], M Chmiel[2], M Brathová[2], S Filipek[1,3]

[1]Institute of Technology and Business in České Budějovice, České Budějovice, Czech Republic, EU
[2]KOVOSVIT MAS Foundry, a.s., Sezimovo Ústí, Czech Republic, EU
[3]University of West Bohemia, Plzeň, Czech Republic, EU

[*]E-mail: svizelova@mail.vstecb.cz

Abstract. Conditions favourable for the development of internal defects in cast pieces may occur in the course of ductile iron casting. The presence of such defects may be fatal for the castings and, together with other factors, result in their disposal. Therefore, any defects should be revealed as early as in the casting design phase if possible. This can be achieved through numerical simulations, a tool commonly used today to predict defects. This paper describes the procedure of numerical modelling of ductile iron casting and solidification of castings intended for the power industry. The purpose of our numerical simulations was to predict filling, solidification and stress states in order to verify whether the proposed production technology is adequate. The numerical simulations revealed some minor deficiencies in the production technology which might lead to defects should an unexpected change in the process conditions occur. In particular, these defects concerned the inlet system, risers, and susceptibility to cracking. Measures to eliminate the risks associated with development of the above-mentioned defects were proposed based on the simulations, which will be integrated in the production technology.

1 Introduction

Nowadays, it is virtually impossible to think of any complex machinery product without a casting. Castings prevail in the structures of fittings, combustion engines, compressors, machining or moulding machines. However, castings can also be encountered in the building industry, architecture, aircraft, shipping, space and nuclear industries [1]. The castings often represent parts whose shapes are characterised by the highest complexity and sometimes even exposed to considerable loads (engine blocks, cylinder heads, pistons and compression rings, etc.). This dictates high demands on quality and an effort to achieve the best possible properties of the castings.

In practice, castings may be porous and include enclosures or show other imperfections, just like any other metallurgical products, which contribute to deteriorated quality. These imperfections are only considered as true defects if they exert an impact on the function or appearance of the product [2]. Porosity is one of the most common defects in castings [3, 4]. In forged pieces, workpieces and other metal products, porosity can be avoided by using intermediate products of adequate quality, by mechanical working itself, etc. [3]. However, porosity in castings not subjected to further processing needs to be addressed by modifying the production technology, preferably before the casting process [4]. The same applies to cracks, fractures and other internal defects that cause fatal impairment of the casting structure integrity.

Numerical simulations are now commonly used in the foundry industry; these simulations can provide many indications of the final casting quality as early as in the design phase. Numerical simulation can be an effective tool for a relatively prompt prediction of changes in the casting structure upon modification of various technological parameters [5-9]. Additionally, they provide a rapid verification of adjustment of the casting geometry, mould dimensions, the inlet system design, risers, vent positions, etc., all this without any financial losses that would be associated with the production of a test casting in the scope of the trial and error method.

The paper presents partial results of research focused on establishing the production of a cast iron with spheroidal graphite as per ČSN EN 1563; the research involved optimization of the production technology of the casting 791 918 SKŘÍŇ 1, representing a component of a gearbox intended for the power industry. Numerous chemical, metallographic and numerical analyses were carried out in the scope of the research, focused on finding feasible ways of minimising defects that occur in the course of the proposed production process of the casting. The following text presents the procedure of numerical modelling of the given casting in MAGMASOFT®, which included calculations of filling, solidification, stress states and deformations. The results were evaluated in respect of identification of a potential risk of internal defects in the casting, particularly porosity and cracks. The determined conclusions will result in optimization of the production technology of the casting 791 918 SKŘÍŇ 1.

2 Casting and material characteristics

The casting identified as 791 918 SKŘÍŇ 1, the subject of numerical simulations, represents the housing component of gearboxes intended for energy industry. Therefore, a high level of material quality is required for this casting to achieve chemical composition, desired structure and mechanical properties. The gross weight of casting is 274 kg and wall thicknesses range between 28-70 mm.

The casting is made of the material EN-GJS-400-18-LT, i.e. cast iron with spheroidal graphite and a ferritic matrix with a guaranteed impact work value at -20°C [10, 11]. Chemical composition of the material EN-GJS-400-18-LT is provided in **table 1**.

The material EN-GJS-400-18-LT offers outstanding machinability, combined with resistance to low temperatures, good fatigue strength, electric conductivity and magnetic permeability, including enhanced ductility and an ability to absorb noise and vibrations well. Mechanical properties of the material EN-GJS-400-18-LT are shown in **table 2**.

Table 1. Chemical composition of the material EN-GJS-400-18-LT

	C	Si	Mn	P	S	Cu	Mg	Cr
Min.	3.50	1.90	0.10	×××	×××	×××	0.040	×××
Max.	3.60	2.00	0.15	0.040	0.010	0.05	0.060	0.02

Table 2. Mechanical properties of the material EN-GJS-400-18-LT (standard wall thickness 30 $<t \leq 60$ mm) [10]

Tensile strength Rm (MPa)	Yield strength $Rp_{0.2}$ (MPa)	Ductility A_5 (%)	Brinell hardness HB (-)	Min. absorbed energy (low temperature (-20 ± 2) °C) KV_2 (J)
Min. 380	Min. 230	Min. 15	Min. 130	12

3 Geometry and input parameters

Geometry of the castings 791 918 SKŘÍŇ 1, together with the inlet system and other components of the casting system (the core, risers, vents, the Y-block, etc.), is shown in **figure 1**. The mold is always formed for 2 pieces of castings. A computational mesh was generated for the casting system geometry. As regards the calculation of stress states in the casting, a constant size of 4 mm of the computational mesh element was chosen for all elements of the casting system (except the mould). The resulting mesh for the casting 791 918 SKŘÍŇ 1, composed of 1,495,998 elements, is presented in **figure 2**.

Figure 1. Castings and gating system geometry

Figure 2. Computational mesh

The input parameters for numerical simulations of the casting 791 918 SKŘÍŇ 1 were defined based on operating data supplied by KOVOSVIT MAS Foundry, a.s. These parameters include chemical composition of the material EN-GJS-400-18-LT, the casting temperature, the inoculation method, materials to make the mould, the casting filling time, etc. Chemical composition of the cast iron is presented in **table 1**. The casting temperature ranged between 1370–1390°C in the operating conditions. The cast iron with spheroidal graphite used to make the casting 791 918 SKŘÍŇ 1 is inoculated in the operating conditions using the pouring method in two phases: pre-inoculation in the ladle and block inoculation in the pouring basin. Single-phase inoculation in the ladle was considered in the calculation given the limits of the MAGMASOFT®.

The calculation of filling, solidification and stress states was carried out in the simulation. Continuous filling of the mould was ensured using the Automatic Filling Control feature that regulated the cast iron flow by controlling the cast iron level in the pouring basin. 1To calculate the stress states, the time of removing the casting from the mould (24 hours) and the time of core shake out from the casting (50 s from casting shake out from the mould) were also defined. In one of the last steps, types of the required results were selected, related to filling, solidification and stress states in the casting. A thus prepared simulation was run and successfully completed. The calculation time was about 3 days.

4 Results analysis

Results of the numerical simulation of filling, solidification and stress states in the casting 791 918 SKŘÍŇ 1 are summarized below. We focused predominantly on risk areas that could potentially lead to defects in the castings.

The casting filling process is shown in **figure 3** together with the temperature field. Clearly, the castings are filled non-uniformly through the inlet system, particularly at the beginning (see **figure 3a**). The casting 2, which is more distant from the pouring basin, is filled later. This discrepancy can be managed by slightly bevelling the runner, which would accelerate the filling of the second casting. Additionally, at the beginning, the molten mass temperature decreases below the liquidus temperature in some areas, particularly in the first casting which is the first to be reached by the metal. A solid phase was detected at this point; however, it melts again in the subsequent course of the filling process. No defects should thus be caused by this phenomenon. A place was identified in the castings where vents should be positioned to prevent any adverse consequences. This part of the casting is indicated in **figure 4**.

IOP Conf. Series: Materials Science and Engineering 1243 (2022) 012008 doi:10.1088/1757-899X/1243/1/012008

a) 20% filled b) 100% filled

Figure 3. Temperature field during filling (°C)

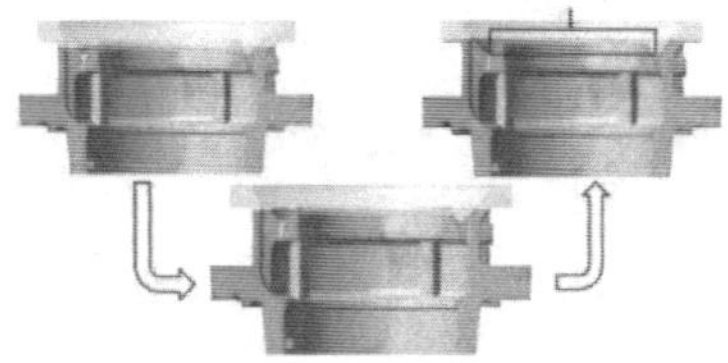

Figure 4. Proposed position of vents

A different view of filling is shown in **figure 5** where flowlines are used to depict the turbulence and whirling during the filling process. Vortices of various dimensions start forming in the volume of individual castings when about one half of them has been filled. Moreover, the metal flowing from the gate falls directly onto the core. This may be a problem for potential erosion of the mould material and of the core. Therefore it would be appropriate to consider modifying the way in which the gates are connected to the casting in order to achieve a more acceptable flow of the molten mass within the casting volume.

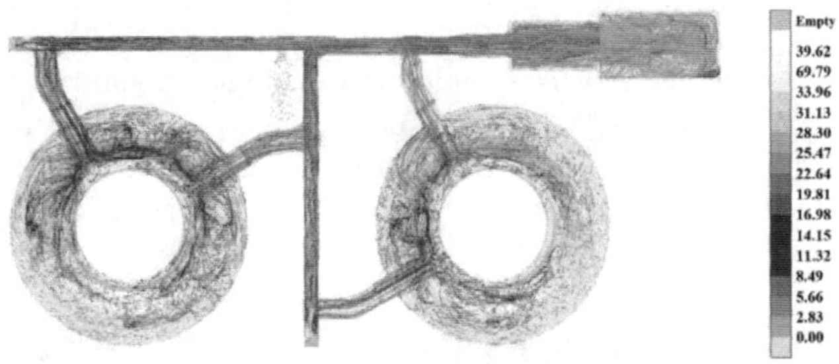

Figure 5. Flow field at the end of filling – Material Age (s)

To give an idea, **figure 6** illustrates the solidification process of Casting 1. As the end of solidification approaches, isolated islets of the liquid phase form, which is undesirable in terms of possible porosity. Temperature nodes exist at the points where closed volumes of the liquid phase are formed as depicted in **figure 6a**. These nodes reach the surface of the casting in some areas. There is also a risk of burning-on at these points. The combination of these two factors indicates a possible presence of sink marks on the casting surface.

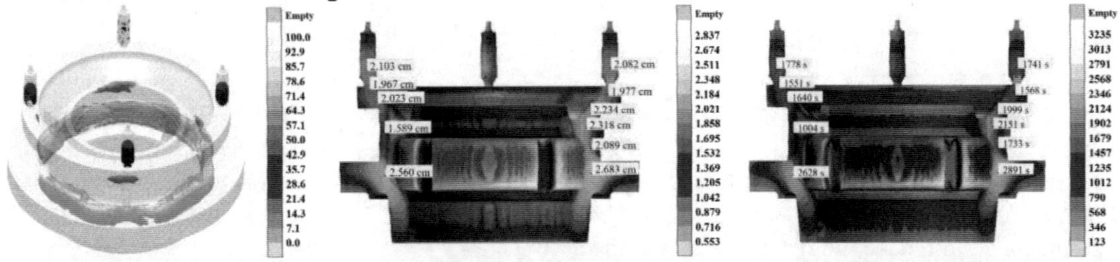

a) Hot Spots (s) b) Feeding Module (cm) c) Solidification Time (s)

Figure 6. Prediction of solidification (Casting 1)

The feeding module is shown in **figure 6b**. As follows from the results, module design of the castings is not optimal. Additionally, the local solidification time (see **figure 6c**) indicates non-directed solidification of the castings with the formation of temperature nodes as mentioned above. These facts may result in porosity in the castings. The porosity should probably be suppressed through graphite expansion; however, if not working properly, the casting porosity will probably be higher than as predicted by the simulation. The situation of Casting 2 was very similar. **Figure 7** shows the prediction of porosity in the castings. The results indicate points where more than 10% of the metal is missing. Porosity may form at these points in the actual casting, as well.

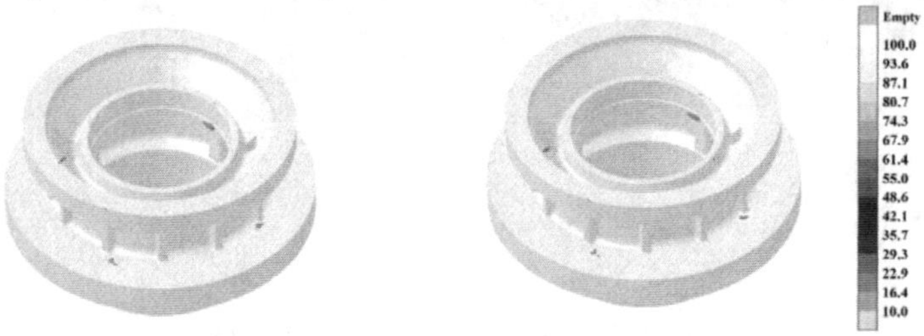

Figure 7. Porosity prediction (%)

Tension accumulates in the casting during the cooling process; this may be associated with a risk if the breaking strength of the material (400 MPa) is exceeded. **figures 8a** and **8b** present a comparison of tensile stress in the casting before and after its removal from the mould. Clearly, a relatively high tensile stress is present in the casting before its removal. This situation poses a risk in terms of cracking. After the casting is removed from the mould, the tension is relieved and the tension state of the casting becomes much more favourable. It is therefore not desirable to keep the casting in the mould for too long, and the time of its retaining in the mould should be optimised (currently, 24 hours). Although the tension is relieved after its removal, places of accumulated tension still remain present in the casting; these points pose a risk in terms of cracking as illustrated in **figure 8c**. In this case, the problem could be resolved by increasing the radius at the problematic edge if allowed by the structural design of the casting.

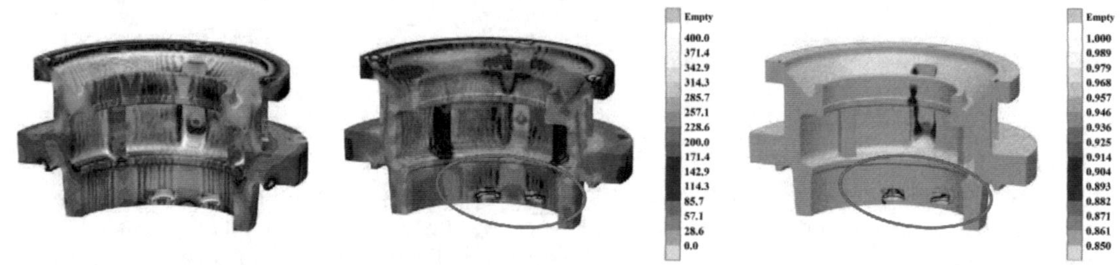

a) Tensile Stress (MPa) – before shake out   b) Tensile Stress (MPa) – after shake out   c) Cold Crack (-)

Figure 8. Hot tearing and cold crack prediction (Casting 1)

5 Conclusions

Numerical analysis of casting the cast piece 791 918 SKŘÍŇ 1 was carried out based on operating data gathered in cooperation with technologists of KOVOSVIT MAS Foundry a.s. The numerical simulation **did not reveal any serious defects** of the casting technology. However, some measures can be recommended to reduce the risk of defects in the castings.

Based on the numerical simulation of the casting 791 918 SKŘÍŇ 1 in the MAGMASOFT®, the conclusions can be summarized as follows:

- Filling of the castings is not uniform. Bevelling of the runner to accelerate filling of the second casting would help to increase filling uniformity. It would be advisable to install vents on the upper flanges of the castings to suppress the risk of insufficient casting. An adjustment of the direction in which the cut is connected to the casting could also be considered; this would achieve more acceptable flow characteristics within the casting volume.

- At the beginning of the filling process, the molten mass cools down below the liquidus temperature and a solid phase develops. The solid phase becomes molten again in the further course of filling. Thus no cold connection defects should appear.
- Risers are not optimal. The casting solidifies in a non-directed manner, which is associated with the formation of unsuitably located temperature nodes. These may lead to porosity in the castings. The porosity should be suppressed through graphite expansion; however, if this does not work, the castings may exhibit higher porosity than that predicted by the simulation.
- Casting tension before and after its removal from the mould does not exceed the tensile strength of the material. However, points where the tension stress approaches the tensile strength occur in the casting before its removal from the mould. If any deviations from the current technological parameters of the production occur, the breaking strength of the material might be exceeded. It is recommended to reduce the casting retaining time in the mould.
- The casting will not have any cracks in the present situation. No fractures will form in the castings. Nevertheless, a point was identified where a risk of cracking is present. The risk can be minimised through an appropriate modification of the casting geometry.

Based on the numerical simulation results, we shall further focus on an optimisation of the casting technology in the operating conditions in order to reduce the risk of defects in the casting 791 918 SKŘÍŇ 1.

Acknowledgements
The paper was prepared under the support of the Czech Ministry of Industry and Trade within the scope of the TRIO programme, as part of project Reg. No. TRIO FV40034 "Research and development of complex technology of castings production from high-quality ductile irons castings". The authors thank the MAGMA team for preparing master studies in MAGMASOFT® for the above project.

References
[1] Stefanescu D M 2017 History of Cast Iron (ASM Handbook: Cast Iron Science and Technology vol 1A) ed D M Stefanescu (ASM Interantional)
[2] Kassie A K and Assfaw S B 2013 Minimization of Casting Defects *IOSR J. Eng.* 3(5) pp 31-38
[3] Monroe R 2005 Porosity in castings. *AFS Trans*, 5(10)
[4] Alagarsamy A and Kumar S 2020 Demystifying Shrinkage Porosity in Ductile Iron Castings *Int. J. Metalcast.* 14(4) pp 1136-1151
[5] Dugic I 2019 Effect of Casting Temperature on the Surface Finish of Grey Iron Castings (*Materials Processing Fundamentals 2019*) ed G Lambotte, J. Lee, A Allanore, S Wagstaff (Cham: Springer International Publishing) pp 87-95
[6] Hawranek R, Lelito J, Suchy J S and Zak P 2009 The Simulation of a Liquid Cast Iron Flow through the Gating System with Filter *Arch. Metal. Mater*. 54(2) pp 351-358
[7] Khan M A A and Sheikh A K 2018. Mechanical characterization and quality of iron castings using optimized mold design: simulations and experimental validation *Int. J. Adv. Manuf. Tech.* 98(1-4) pp 799-809
[8] Sonne M R, Frandsen J O and Hattel J H 2015 Comparison of residual stresses in sand- and chill casting of ductile cast iron wind turbine main shafts *IOP C. Ser-Mat. Sci* 84
[9] Sheikh A K, Khan M A A, Iqbal H and Al-Shaer B S 2018 Casting of adjuster bracket—process optimization and validation *Mater. Manuf. Process*. 33(16) pp 1845-1850
[10] ČSN EN 1563. Founding – Spheroidal graphite cast irons (Prague: The Czech office for standards, metrology and testing) p. 40.
[11] ČSN EN ISO 945-1. Microstructure of cast irons - Part 1: Graphite classification by visual analysis (Prague: The Czech office for standards, metrology and testing) p 68

IOP Conf. Series: Materials Science and Engineering 1243 (2022) 012009 doi:10.1088/1757-899X/1243/1/012009

Testing of wear properties of inserts made from semi-solid processed experimental steel

Opatová K[1], Zatloukal T[2], Jirková[1] H and Volkmanová J[1]

[1] University of West Bohemia, RTI - Regional Technological Institute, Univerzitní 8, CZ-301 00 Pilsen, Czech Republic

[2] University of West Bohemia, Department of machining technology, Univerzitní 8, CZ-301 00 Pilsen, Czech Republic

*E-mail: opatovak@rti.zcu.cz

Abstract. To reduce the cost of manufacturing tools, which are commonly made from powder steels, high-alloy tool steel in the form of an ingot was cast. Conventional casting allows better use of recycled materials and an overall reduction in material production costs. This experimental steel was then processed in a semi-solid state to remove the unsuitable casting structure and to obtain the desired functional properties. In the case of tools, it is primarily a sufficient toughness and abrasion resistance. To determine the wear resistance of the experimental steel, replaceable inserts for machining tests were made. The inserts were prepared by electro-erosive machining. During testing, the geometry of the insert was optimized to achieve good chip evacuation and other cutting properties. Wear was measured using a digital microscope and at the same time, the roughness of the workpiece was measured. For comparison, a powder steel insert of the corresponding chemical composition was also produced, which has the commercial title Böhler K390 Microclean. This steel was heat-treated according to the manufacturer's recommendations. By comparing the results obtained on experimental cast steel and commercially produced powder steel, it was found that cast steel processed by semi-solid technology achieves three times lower cutting-edge wear during machining than powder steel.

1 Introduction

The development of tools for both machining and forming has undergone significant development. Progress has been made both in the development of tool steels with a high content of alloying elements and in the field of powder steels. Steels produced by powder metallurgy are currently very often used due to their cutting properties and wear resistance. However, the technology of powder metallurgy is very time-consuming, energy-intensive and expensive, and therefore new possibilities are constantly being investigated to develop a material or processing method that could achieve similar results. Such a route could be the processing of materials in the semi-solid state. For that purpose, the processing of conventionally casted high-alloy steel in semi-solid state has been investigated in this work [1]–[4], [10].

For the experimental program, high-alloy tool steel with corresponding chemical composition as commercially produced cold work powder steel Böhler K390 was cast. This steel is characterized by good mechanical properties, especially high wear resistance, hardness and compressive strength. This is due to its alloying, where vanadium of which 9 wt % occurs in steel predominates. Other important alloying elements are carbon, molybdenum, cobalt, tungsten and chromium (table 1). The combination of alloying elements predeterminate this material for use as forming tools for cutting and bending. Such alloyed materials are usually produced by powder metallurgy. In this way, parts of the final geometry are obtained, which are ready for operation after heat treatment. The heat treatment of such steels is similar to the processing of tool steels. The advantage of powder metallurgy is the low proportion of waste material, but a significant disadvantage is the financial and energy consumption during the production process. Therefore, the experimental material marked KK steel was cast in a conventional way to reduce the material costs. On the other hand, high-alloy cast steels form an unsuitable casting structure with large primary carbides that are difficult to eliminate. For this reason, the experimental KK steel was heated up to semi-solid state, i.e. at temperatures at which these primary carbides can be partially eliminated [5], [6][7].

Table 1. Chemical compostion of the conventionally casted experimental KK steel and the commercially used powder steel Böhle K390 microclean

Chemical comp. [wt%]	C	Si	Mn	P	S	Cr	Mo	Cu	Co	V	W
Experimental steel KK	2.49	0.61	0.29	0.016	0.005	4.21	3.71	0.13	2.11	9.12	0.97
Böhler K390 Microclean	2.47	0.55	0.40	-	-	4.20	3.80	-	2.00	9.00	1.00

2 Experimental program

For the design of thermal and thermo-mechanical processing of experimental steel KK and especially for its processing in the semi-solid state, it was necessary to determine the dependence of the liquid phase content on temperature. For this purpose, a graph of changes in the phase composition as a function of temperature was calculated in the software JMatPro (SENTE SOFTWARE LTD. JMatPro, Release 9.0. 2016). This was used to determine the temperature range when the proportion of liquid phase in the structure corresponds to about 15 - 35%. The calculated temperature range was 1260 - 1290 °C (figure 1). For the experiment, the interval was extended to 1220 - 1300 °C, as the calculated values were only an approximation of the real state. In this temperature range, the determination of the ideal processing temperature of KK steel in the semi-solid state was then carried out. Direct experimental determination of an ideal temperature without software support would be too material, time and financially demanding [1], [3], [8].

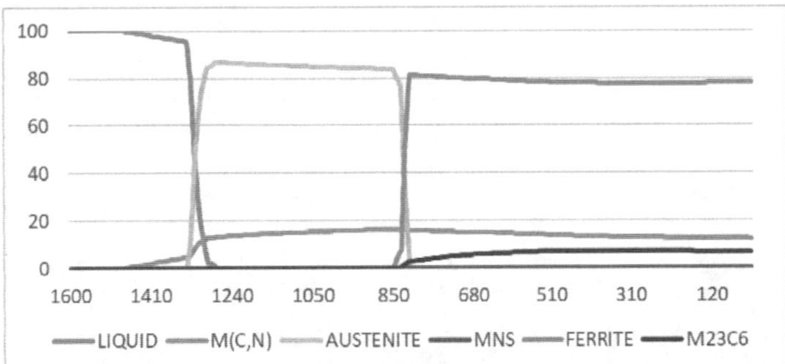

Figure 1. The dependence of phases volume of the experimental material on the temperature

When the optimal semi-solid temperature, where approximately 30% of the liquid phase is present in the material during processing, was found, the next step took place. It was necessary to optimize the whole process of forming. It was found that the optimal regime consists of heating the material to the semi-solid temperature 1260°C with a holding time of 20 minutes in order to partially dissolve the primary carbides which were in the casted structure. Then the material was quenched for 3 seconds in the water and the holding on the temperature of 1100°C followed. Soak at this temperature was either followed by the insertion of a single or double deformation, or direct quenching into the

IOP Conf. Series: Materials Science and Engineering 1243 (2022) 012009 doi:10.1088/1757-899X/1243/1/012009

water without the insertion of a deformation. After the semi-solid treatment, the heat treatment was investigated, which, if used, consisted of double tempering for two hours at a temperature of 500° C. To evaluate the properties of the experimental steel which was obtained by the semi-solid processing, indexable inserts for turning were made from four different states of the formed material. For the possibility of comparison with the conventional powder material, an additional insert from powder steel Böhler K390 microclean was made (figure 2).

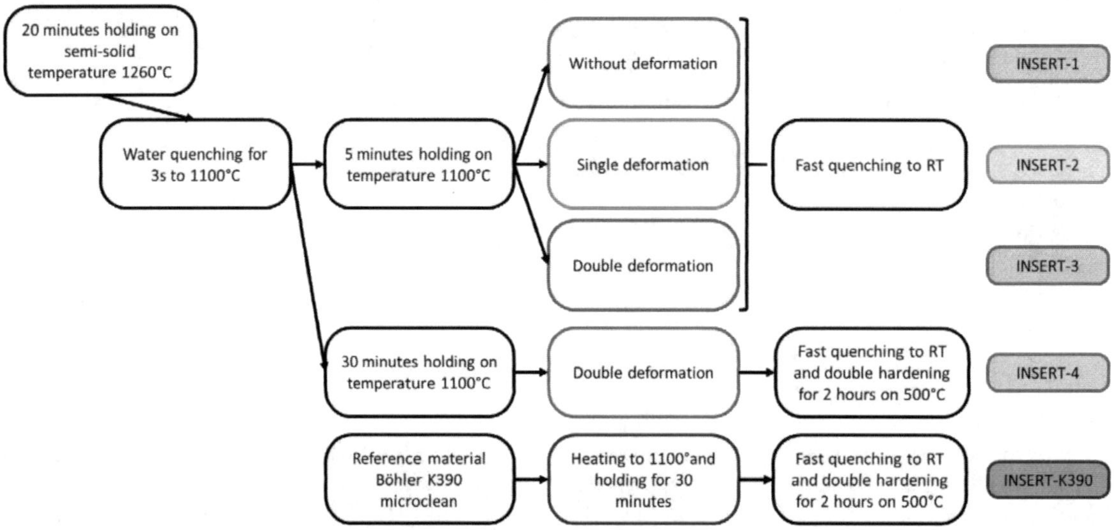

Figure 2. The scheme of experimental material processing routes and inserts manufacturing

2.1 *Manufacturing of the inserts*

Replaceable experimental high-alloy steel inserts were made by cutting a semi-finished product from a semi-solid-formed material by electro-erosion machining (EDM). The chip breaker is usually made using a tool grinder, but it can also be obtained using an unconventional method of electro erosive machining, as was the case with the tested tools. Using electro erosive machining, it was possible to create complex geometries with a single device. The four-axis arrangement of the EDM was also suitable for the production of complex chip breaker shapes of replaceable inserts. The advantage of electro-erosive machining over the use of a tool grinder is the possibility of a smooth transition of the cutter shape from narrower to wider. This is not possible when using one grinding element. Utilizing the advantages of EDM, CNMG inserts with an original breaker geometry were produced (figure 3) [9].

Figure 3. Electro-erosive machine and insert made from experimental material by electro-erosive machining

3 Results and discussion

In the first machining step, the optimal machining parameters suitable for this type of insert material were found. The experimental steel had no further surface treatment (TiN thin layer, etc.) and therefore the parameters were similar to those used for tools made of high-speed steel. Subsequently, all inserts were tested under the same conditions. The machined workpiece was a cylinder from C45 steel with an initial diameter of 98 mm and a length of 400 mm. The cutting speed v_c was 45 m/min, the feed f was 0.15 mm and the depth of cut a_p was 1 mm. During the wear test, either the machining time or, as in this case, the cutting length is selected as the next constant. This was 20 mm for each step monitored. After each step, the depth of wear and the amount of built-up edge (BUE) were measured on a Keyence VHX-6000 digital microscope. A total of 15 sections were tested (300 mm of cut in total). At the same time, the roughness of the workpiece was determined after each cut using a movable Taylor Hopson - Surtronic Duo Surface roughness tester.

Inserts that were not tempered at 500 ° C during manufacturing showed worse wear resistance than an INSERT-4 that was tempered twice at 500 °C. The INSERT-1, which was not deformed, was only heated to a semi-solid temperature with a holding time of 20 min, then quenched for 3 s in water, heated to 1100 ° C and quenched in water, showed significant v_{BN} wear and constant increase of built-up edge. The wear of the flank increased evenly during testing up to a value of 0.22 mm. The increase varied unevenly during testing and the roughness of the machined surface, which was measured by a portable roughness meter, increased unevenly during testing up to R_a = 4.4 μm (figure 4).

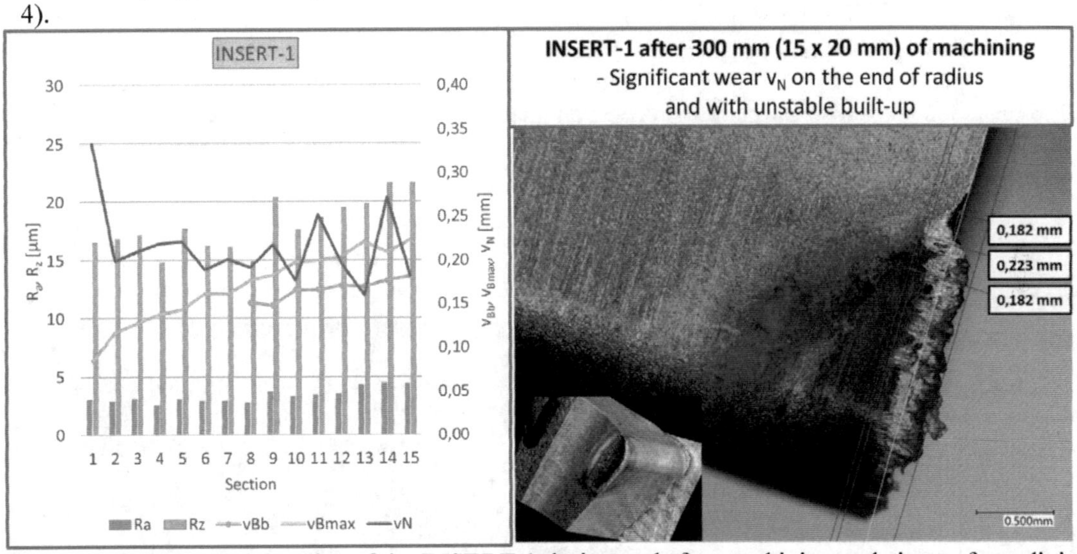

Figure 4. Functional properties of the INSERT-1 during and after machining and picture from digital microscope with built-up edge (BUE) and significant wear

The INSERT-2 was processed in a regime in which the material was heated to a semi-solid temperature of 1260 ° C, then quenched in water for 3 s, heated to a forging temperature of 1100C, where it was subjected to single deformation and then water quenched, which resulted in fine microstructure with martensitic matrix. The specimen was not subjected to subsequent tempering which led to tempering during tensting. This method of processing resulted in an insert that had a wear v_{Bmax} = 0.22 mm after the fifteenth cut. This was the same value as for the INSERT-1, the values of the workpiece surface roughness were similar, which remained practically the same during testing, around R_a = 18 μm. No significant plastic deformation was observed on the insert. However, there was significant wear v_{BN} on the end of the chip outcome and a hole in the chip breaker caused by chips departure (figure 5).

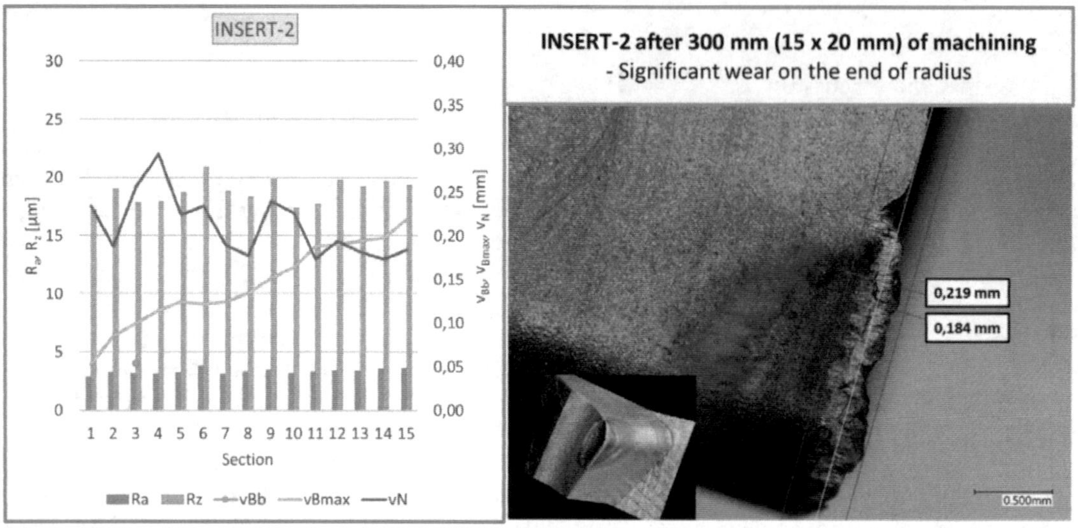

Figure 5. Functional properties of the INSERT-2 during and after machining and picture from digital microscope which showing the surface affected by high temperatures during machining

INSERT-3, which has been processed in the same way as INSERT-1 in addition, it was subjected to double deformation, i.e. it was heated to a semi-solid temperature of 1260 °C with a holding time of 20 min. It was then quenched in water for 3 seconds, heated to 1100 °C for 5 minutes and then formed on a hydraulic press and quenched in water. Since no tempering was performed at 500 ° C before testing, the insert material was tempered during testing. This led to plastic deformation at the tip and cutting edge of the insert, which was visible on images from the Keyence digital microscope. Plastic deformation of the insert and the formation of a relatively stable built-up edge led to wear v_{Bmax} = 0.26 mm after the fifteenth cut. Behind the reinforced part of the insert edge, there was again significant wear v_{BN} (Figure 6).

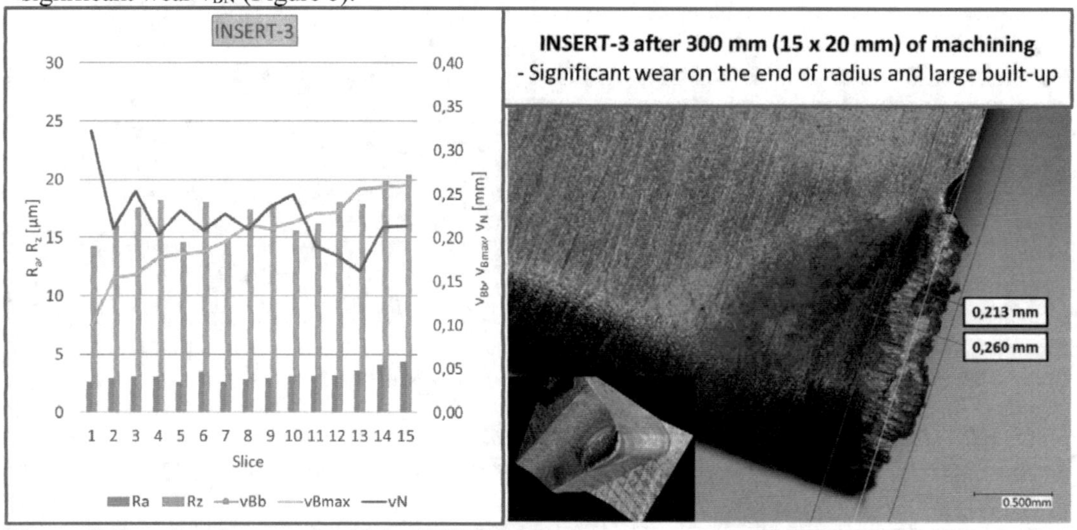

Figure 6. Functional properties of the INSERT-3 during and after machining and picture from digital microscope showing the plastic deformation of the tip of the insert

The INSERT-4, which was processed in a complex mode, achieved the best results in wear tests. First, it was heated to a semi-solid temperature of 1260 °C, followed by quenching for 3 s in water, heating to a forging temperature of 1100 °C, where an extended holding time of 30 min was performed before double deformation. The deformation was followed by quenching in water and double tempering for 2 hours at a temperature of 500 ° C. The insert made of the material treated in this way achieved wear of v_{Bmax} of only 0.19 mm after 15 cuts with a length of 20 mm, while two types of wear were distinguished during the measurement. The maximum wear was measured at the end of the cutting edge, due to the reinforced layer after previous turning and the average wear v_{Bc} was measured on the linear part of the cutting edge. This reached only 0.095 mm after fifteen cuts. On the other hand, worse values were achieved when measuring the roughness of the machined part. It gradually increased to R_a = 4.8 μm, but the initial roughness was the lowest of all tests, namely R_a = 2.3 μm (figure 7).

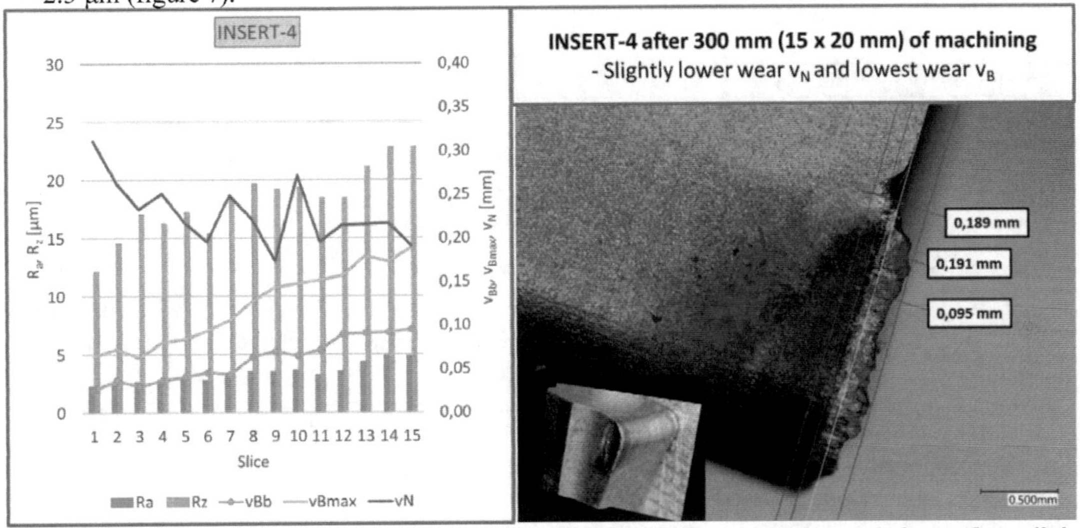

Figure 7. Functional properties of the INSERT-4 during and after machining and picture from digital microscope with no signs of plastic deformation and only small built-up edge

The INSERT-K390, made of Böhler K390 microclean reference powder steel and treated by quenching from 1100 ° C and tempering at 500 ° C for two hours, achieved a similar average v_{Bc} wear as the INSERT-4, specifically worn to a depth of 0.21 mm after 15 cuts. However, it had significantly greater maximum wear, 0.37 mm, and also achieved significantly worse values of the surface roughness of the machined surface than all previous inserts. The final workpiece roughness after wear testing was R_a = 6.7 μm (figure 8).

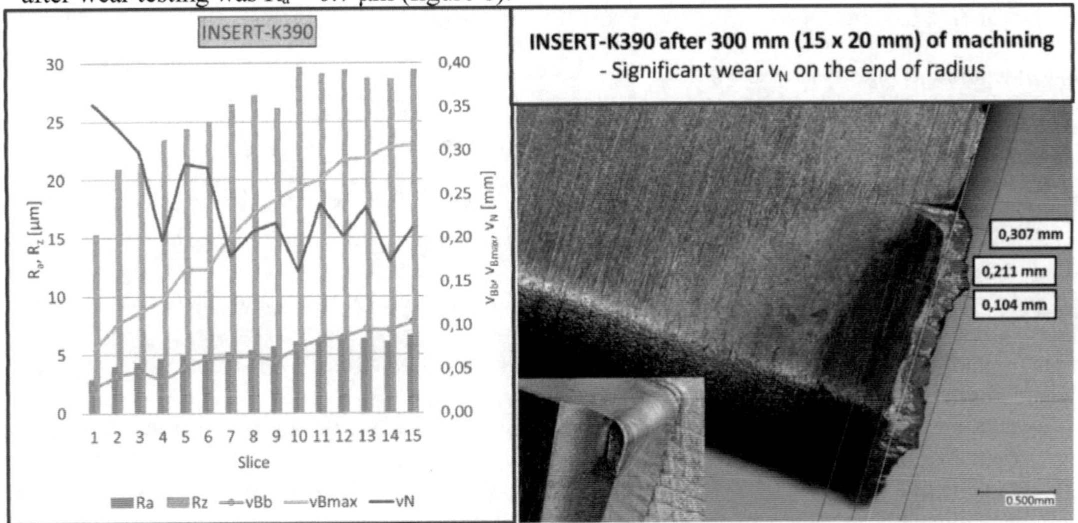

Figure 8. Functional properties of the INSERT-1 during and after machining and picture from digital microscope

Overall, the INSERT-4 showed the best wear and machining behavior, having mean wear of only 0.095 mm after wear tests. That indicated that using these machining parameters would last the longest of the inserts in the cut while maintaining the desired workpiece surface quality (table 2).

Table 2. The overall results from the wear tests showing the potential of inserts made by semi-solid forming from casted high-alloy steel

Insert	Max wear [mm]	Average wear [mm]	Average roughness R_a [µm]	Average hardness VH0,1	Testing parameters
INSERT-1	0,223	-	3,372	897	v_c = 45 m
INSERT-2	0,219	-	3,338	**903**	f = 0,15 mm
INSERT-3	0,26	-	**3,155**	852	a_p = 2 mm
INSERT-4	**0,189**	**0,095**	3,414	827	machined trajectory:
INSERT-K390	0,307	0,104	5,31	812	15x20 mm

4 Conclusion

In order to reduce the cost of production of materials with high wear resistance, which are usually produced by powder metallurgy, high-alloy tool steel was conventionally cast. It had a chemical composition corresponding to commercially produced Böhler K390 Microclean powder steel. The conventional casting of materials with a proportion of alloying elements of about 20 wt% is usually not performed due to the unsuitable casting structure which these materials form during cooling. These are large sharp-edged carbides and carbide networks. These structural elements significantly impair the mechanical properties of the material and are very difficult to remove by conventional forming and heat treatment methods. By optimizing the parameters of unconventional processing of cast high-alloy tool steel, the material with process properties corresponding to steel produced by powder metallurgy was obtained. Inserts were made from the processed material to determine the functional properties of experimental steel. For comparison, the insert was also made of Böhler K390 microclean. All inserts were tested under the same machining parameters, with the INSERT-4 showing the best properties after machining a 300 mm along the length of the workpiece. INSERT-4 was processed by a complex regime consisting of heating to semi-solid temperature with holding, cooling to forging temperature, insertion of double deformation, quenching to RT and subsequent tempering. The insert processed in this way achieved the lowest wear of all samples, specifically v_{Bmax} = 0.189 mm. The wear was about a third lower than the maximum wear v_{Bmax} of the powder steel insert.

Acknowledgements

This article was created under the project SGS-2019-008: Research and Development for Innovation in the Field of Manufacturing Technology - Machining Technology III.

References

[1] George E. Totten 2006 Steel Heat Treatment Metallurgy and Technologies Second edi. (Boca Raton: Taylor & Francis)
[2] George E. Totten and Rafael Colás 2016 Encyclopedia of Iron, Steel, and Their Alloys 1st editio (CRC Press)
[3] Attaullah Arain 1999 Heat Treatment and Toughness Behavior of Tool Steels (D2 and H13) for Cutting Blades (University of Toronto) p 1–80
[4] Banerjee M K 2017 2.8. Heat Treatment of Commercial Steels for Engineering Applications (Elsevier Ltd.)
[5] Mazancová E 2012 Technické materiály I. Kovové materiály (Ostrava: Vysoká škola Báňská - Technická univerzita Ostrava)
[6] Kianian B 2019 Comparing acquisition and operation life cycle costs of powder metallurgy and

conventional wrought steel gear manufacturing techniques *Procedia CIRP* **81** p 1101–6
[7] Yan J, Zhou Y, Gu R, Zhang X, Quach W M, and Yan M 2019 A comprehensive study of steel powders (316L, H13, P20 and 18Ni300) for their selective laser melting additive manufacturing *Metals (Basel)* **9**
[8] Jurči P 2009 Nedostatky při výběru a zpracování vysokolegovaných nástrojových ocelí *Metal conference*
[9] Mohd Abbas N, Solomon D. G., and Fuad Bahari M 2007 A review on current research trends in electrical discharge machining (EDM) *Int. J. Mach. Tools Manuf.* **47** p 1214–28
[10] Rubešová K, Jenicek Š, Kana, J, Zetková, I 2016. Microstructure of MS1 Maraging Steel in 3D-Printed Products After Semi-Solid Processing. 10.2507/27th.daaam.proceedings.070.

IOP Conf. Series: Materials Science and Engineering 1243 (2022) 012010 doi:10.1088/1757-899X/1243/1/012010

Usage of the 3D printed model for the forming during the development of the casting mould

Jiri Korinek[1*], Karel Raz[1], Zdenek Chval[1]

[1]University of West Bohemia, Regional Technological Institute, Univerzitni 8, 301 00, Pilsen, Czech Republic

[*]E-mail: jkorinek@rti.zcu.cz

Abstract. This paper deals with the usage of the plastic model (created by additive technology), which can be used for the mould production as an internal model for the cavity. This mould is intended for the steel casting technology. The particular plastic model is created by the conventional 3D printer HP Multi Jet Fusion 4200 from the material PA 12 GB. It is polyamide with 40% of glass filling. This material is suitable for this method because of good mechanical properties. The creating of the steel casting mould has some specification which are deeply described in this paper. The comparing of several dimensions between the CAD model, plastic model and the final casted model from steel is performed here. It is obvious that the differences and tolerances are less than 1%. This approach can be used as a replacement of the wood models which are generally used. The benefits of the 3D printed model are higher accuracy of the model, higher complexity of the model and production time which can be less than 3 days (from the CAD model to the plastic model for the casting mould, but depending on complexity and size of the model).

1. Introduction

The area of the usage of products which are produced by the technology of classical casting from cast irons has nowadays greatly decreased due to the new technological and design possibilities. It is caused by more available technology of cutting of flat parts (sheets and profiles) with the of laser or plasma. It is generally possible to replace casted parts by precise and complex weldments. It is caused by higher technological, time and financial demands on the production of steel castings. This aspect is resulting in a small number of foundries, which are able to produce parts with more complex shapes. We are considering foundries, which are dealing with the production of castings weighing up to about 500 kg. [1,2]

Actual problem with casting is a lower number of highly qualified workers. Nowadays is still decreasing the number of people which are interested in crafts.

The creation of a model, for example, for a casting into a sand mold is one of the most important steps within the chain of the steel castings production. The classical approach is the creation of a model from wood, by wood-modelers and by usage of classical operations - using machining, or for more complex shaped parts using handwork, where the skill of the wood-modeler had to be reflected. [3,4]

An important aspect is also the experience of the modeler in the creation of individual shapes in relation not only to the creation of the mold, but also to the casting itself and the technology of steel casting. [5]

At present, it is possible to use the 3D printing technology for the model production. It can highly increase accuracy of the model, decrease the delivering time of the model and decrease the costs of the final product. [6]

2. Description of the part and inputs

The inlet cone for the water turbine (Kaplan type) was selected as the testing part in this research. This part is produced by casting from the cast iron. This turbine is a machine with vertical arrangement with an impeller diameter of 470mm. The detail and technical description of the inlet cone is obvious from the following figure (Figure 1). [7]

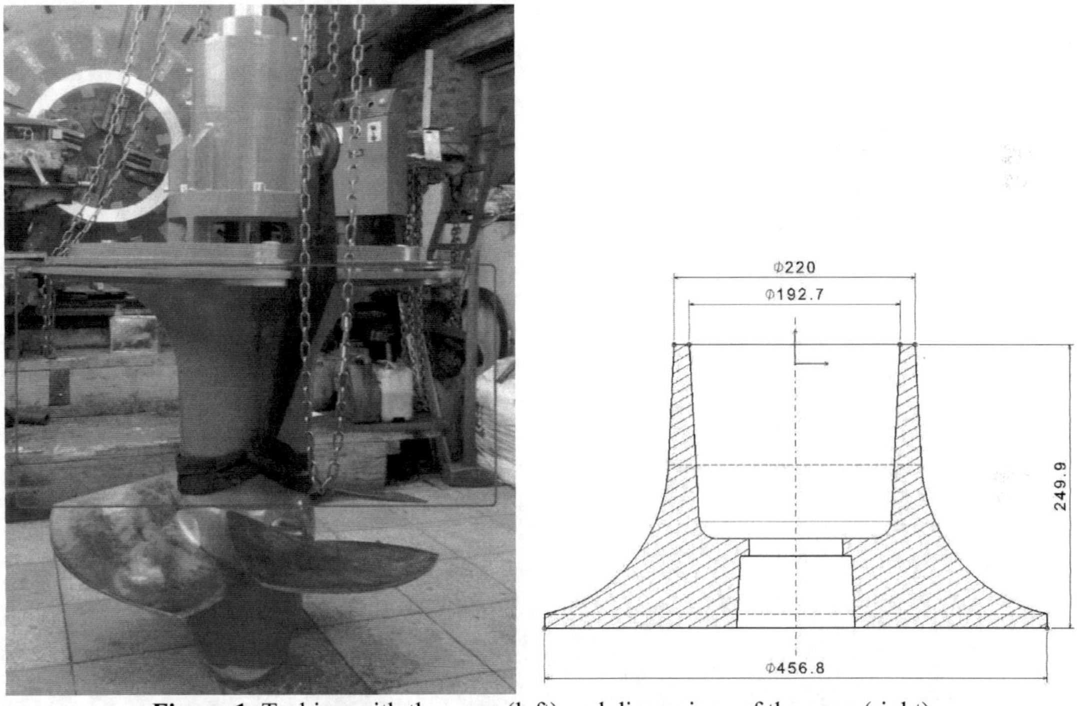

Figure 1. Turbine with the cone (left) and dimensions of the cone (right)

Casting from the gray cast iron is the only possibility for the production of this part is (because of the shape of this cone). This part is used as special segment for directing the inflow of water to the impeller [8]. There is not a significant loading of this part during the operation life. It is necessary to consider the long-term usage and the effect of cavitation and abrasion caused by the sand and other impurities in the water. [9,10]

2.1. Creation of the 3D CAD model

The 3D CAD model was creating according the previously destroyed part. There was added 5mm allowance on the outer surface for the machining and the whole model was enlarged by 3% because of the shrinkage. This value was to large after the first casting (it was not necessary too much) and the final value of the volumetric shrinkage was about 1%. [11]

3. Usage of the additive technology during the model production

The 3D CAD model of the final casting model has to be upgraded with respect to the used additive technology. The part was divided into several parts due to the size of the model and the printing space of the printer. A division into quarters was chosen. There was shape connection (hole and pin) on the contact surface in order to ensure the precise position of each part. Individual parts were connected by M6 bolts. Another modification of the model was the creation of a shell body with a wall thickness of 3 mm. This step was used because of plastic material savings. This part is shown in Figure 2. [12, 13]

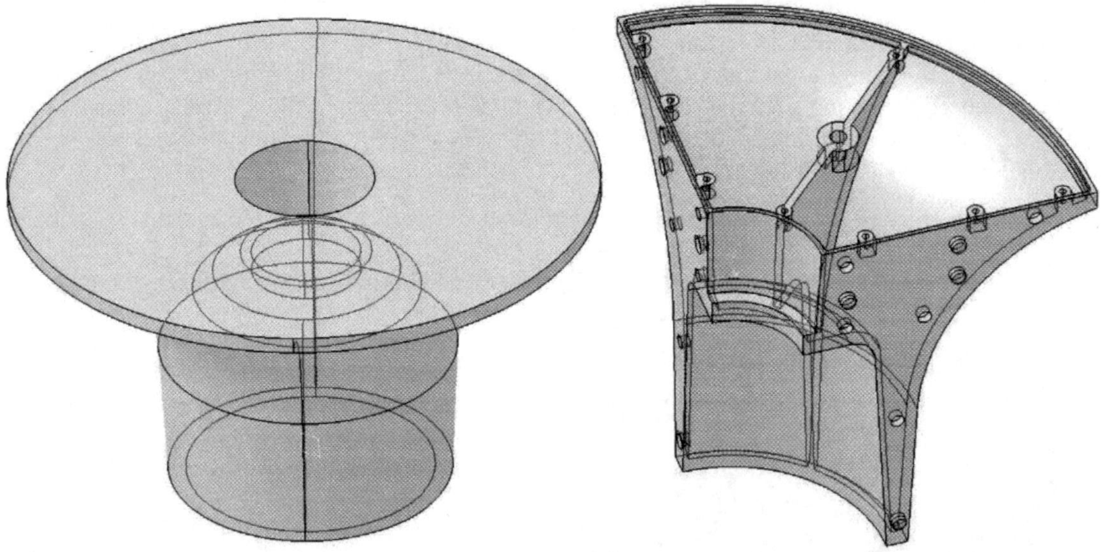

Figure 2. 3D CAD model of the cone (left), one quarter of the model after design changes (right)

It was necessary to make the model open (not closed spaces can be used) due to the removal of unused printing material. The internal cavity was closed by the cover. Some ribs were added to the design in order to improve the stiffness of the plastic model. An internal thread was added to the model in order the remove this model from the sand mould after the moulding process by usage of this threaded hole (Figure 3).

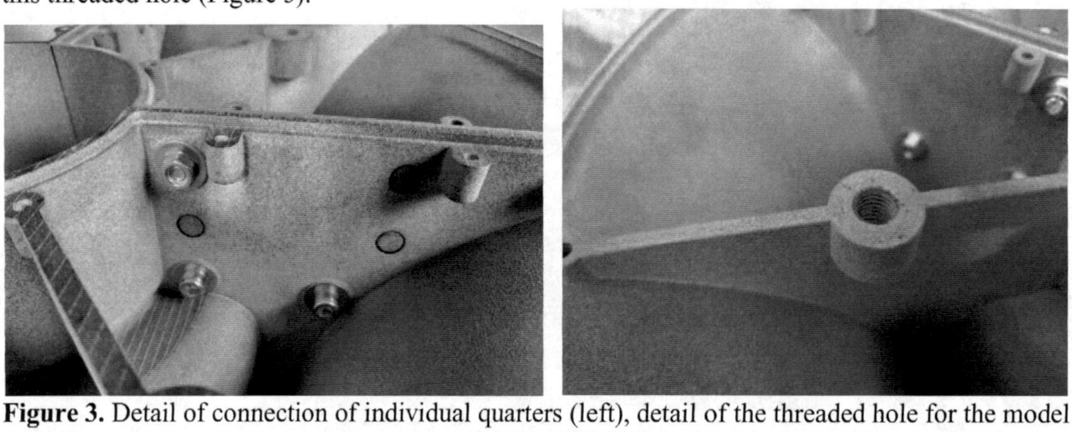

Figure 3. Detail of connection of individual quarters (left), detail of the threaded hole for the model removing from mould (right)

The whole model was finished by the two-component polyurethane mastic sealant after assembling of all parts was together. Special polyurethane painting was used after this process in order to make the surface smooth and decrease the roughness. It can be seen in the Figure 4.

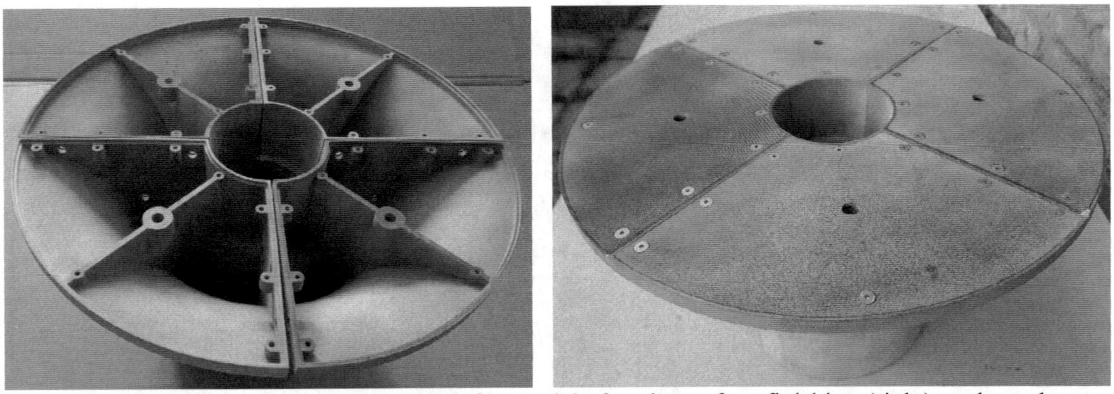

Figure 4. Model after the 3D print (left), model after the surface finishing (right)- polyurethane mastic sealant is only on functional surfaces (yellow color)

3.1. Description of the additive technology

The printer HP Multi Jet Fusion 4200 (Figure 5) was used for the production of the model. It is a very precise and productive industrially usable 3D printer, which is working on the principle of sintering of plastic powder, placed in a container. This printer does not need any supports and it saves material by this technology. The part from this printer can be used immediately after the printing process without any special post-processing (such as machining or supports removing). The mechanical properties of the resulting product can be considered in all 3 directions as the isotropic material. This is big advantage comparing to other printers, especially comparing with FDM printers, which is one of their main disadvantages. [14]

The printing space of the own printer 380 x 380 x284 mm and the printed material is PA12GB, which is a polyamide with an 40 % of glass contain. The actual printing time of the model is less than 10 hours. After the actual printing, it is necessary to perform cooling and postprocessing (cleaning). It can be said that the final model can be obtained within 3 days from the beginning of printing. [15,16]

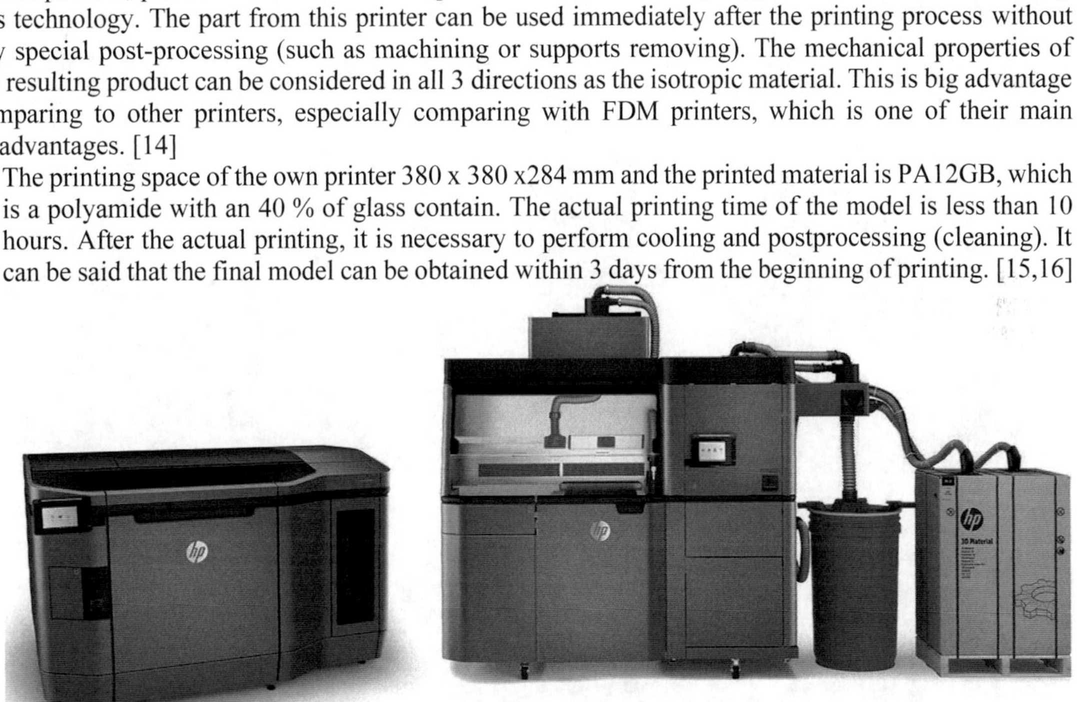

Figure 5. The printer HP Multi Jet Fusion 4200 with post-processing kit

4. Casting of the final part from the cast iron

The casted part from iron (Figure 6) was completed in two weeks from the start of the production. The weight of the casting was 54 kg. The casting itself did not show any surface defects.

According to the information from the foundry, everything was performed according to the established procedures and assumptions, and the model was therefore fully in line with the traditional wooden model.

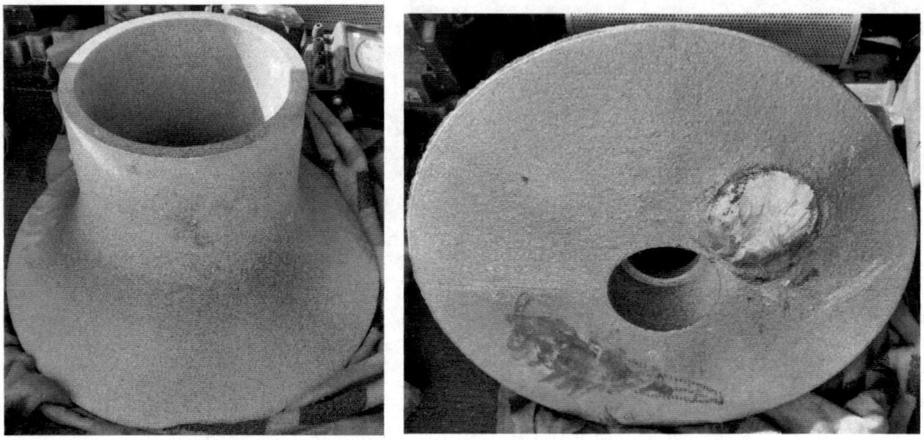

Figure 6. Full view on the final part from the cast iron

5. Validation of the geometrical dimensions of the casted part with respect to the 3D CAD model and the 3D printed model

The comparison of the required geometric dimension with the real values is an interesting parameter od this research. Four basic dimensions were measured on the model, which are shown in the following figures (Figure 7).

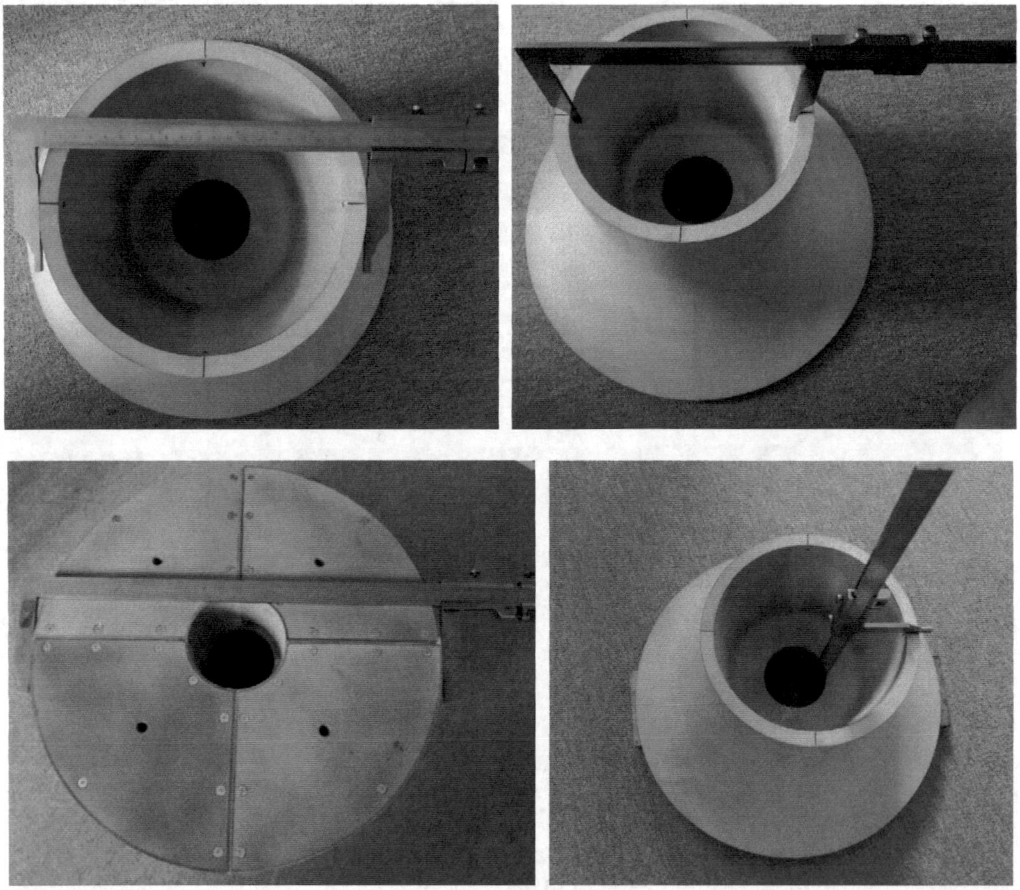

Figure 7. Dimensions for comparing: A (upper left), B (upper right), C (bottom left), D (bottom right)

The measurement results, description of values and comparisons of individual values are shown in the Table 1.

Table 1. Dimensions comparing

Dimension description	Dimension label	CAD model [mm]	3D print [mm]	Casted part [mm]	Difference CAD model X 3D print [mm]	Difference CAD model X Casting [mm]
Outer diameter at outlet	A	220	220.9	220.5	+0.9	+0.5
Inner diameter at outlet	B	192.7	191.8	191.5	-0.9	-1.2
Outer diameter of basement	C	456.8	458.7	457.3	+1.9	-0.5
Total height	D	249.9	250.7	249.1	+0.8	-0.8

It is obvious (from the table above), that dimension of the 3D CAD model, 3D printed part and the final casted part are within considered tolerances. It is necessary to notice, that the total difference of each individual dimension didn't exceed 1%.

6. Conclusion

According to the available information from the foundry, the costs of casting by usage of the model from a 3D printer and using the classical model (wooden model) are practically the same. It has to be noticed, that this is comparing performed for this particular model of turbine cone. It should be noted that this is a relatively simple shape, which can be relatively simply produced by classical methods of machining.

The printing technology will be much more productive for parts with more complex shapes. The repeatability of model production is another big advantage of this technology.

3D printed models can be easily used also for inlet system models. The simulation and optimization of the inlet system can be used here without any problem.

The most questionable parameter is the time. Everything depends on the capacity of the printer. This research showed big advantage of this method, which can again improve the usability of casted products. It is caused by the relatively simple method of the model production. The research will continue with the production of more complex part and economical evaluation of the price of the iron-casted product (considering the model for forming from wood and from plastic material).

Acknowledgement

The present contribution was supported from the project ERDF "Research of additive technologies for future applications in the machinery industry - RTI plus" (No. CZ.02.1.01/0.0/0.0/18_069/0010040).

References

[1] F. Ettenmeyer, M Schweinefuss, P.Lechner, J. Stahl, T. Gress, J. Kaindl, L. Durach, W. Volk. *Characterisation of the deoring behaviour of inorganically bound cast-in cores for light metal casting*, Journal of Materials Processing Technology, 2021, https://doi.org/10.1016/j.jmatprotec.2021.117201

[2] E. A.Chernyshov, A. D. Romanov, E. A. Romanova. *The quality control of high resistance steel castings by optimizing the tempering temperrature*, Materials Today, ICMTMTE 2020, Sevastopol, DOI: 10.1016/j.matpr.2020.08.134

[3] S. Yu, M. Long, M. Zhang, D. Chen, P. Xu, H. Duan, J. Yang. *Effect of mold corner structures on the fluid flow, heat transfer and inclusion motion in slab continuous casting molds* , Journal

of Manufacturing Processes, Volume 68, pp. 1748-1802, 2021, DOI: 10.1016/j.jmapro.2021.06.067

[4] C. P. Egole, H. E. Mgbemere, G. M. Sobamowo, G. I. Lawal. *Micro-macro model for the transient heat and transport in solidification structure evolution during static casting process*, Materials Today Comunications, 2021, DOI: 10.1016/j.mtcomm.2021.102613

[5] Alketan, Oraib & Rowshan, Reza & Abu Al-Rub, Rashid. (2018). *Topology-Mechanical Property Relationship of 3D Printed Strut, Skeletal, and Sheet Based Periodic Metallic Cellular Materials*. Additive Manufacturing. 19. 167–183. 10.1016/j.addma.2017.12.006.

[6] O' Connor, Heather & Dickson, Andrew & Dowling, Denis. (2018). *Evaluation of the mechanical performance of polymer parts fabricated using a production scale multi jet fusion printing process*. Additive Manufacturing. 22. 10.1016/j.addma.2018.05.035.

[7] K. Omer, A. Abolhasani, S. Kim, T. Nikdejad, C. Butcher, M. Wells, S. Esmaeili and M. Worswick, "Process parameters for hot stamping of AA7075 and D-7xxx to achieve high performance aged products," *Journal of Materials Processing Technology*, vol. 257, pp. 170-179, 2018.

[8] Voyiadjis GZ, Woelke PB, 2005 "*General non-linear finite element analysis of thick plates and shells*". Int. J. Solids Struct, 43 (7–8, 1):2209–42

[9] N. Arcuri, F. Reda, M. De Simone, *Energy and thermo-fluid-dynamics evaluations of photovoltaic panels cooled by water and air*, Solar Energy 105 (2014) 147–156

[10] Z. Chval, K. Raz, Effect of heat load on a mechanical forging press. In Proceedings of the 27th DAAAM International Symposium, 2016, Vienna: DAAAM International, pp. 344-348. ISBN: 978-3-902734-08-2, ISSN: 1726-9679

[11] V. Marek, *Basic Research of Thermal Transfer Simulations*, Proceedings of the 27th DAAAM International Symposium, pp.0578-0585, B. Katalinic (Ed.), Published by DAAAM International, 2016, ISBN 978-3-902734-08-2, ISSN 1726-9679, Vienna, Austria, DOI: 10.2507/27th.daaam.proceedings.085

[12] V. Marek, Z. Hajicek, *Thermal Simulations Based on Macro-Models*, Proceedings of the 28th DAAAM International Symposium, pp.0627-0634, B. Katalinic (Ed.), Published by DAAAM International, 2017, ISBN 978-3-902734-11-2, ISSN 1726-9679, Vienna, Austria, DOI: 10.2507/28th.daaam.proceedings.089

[13] Gorguluarslan, Recep & Gandhi, Umesh & Mandapati, Raghuram & Choi, Seung-Kyum. (2015). *Design and fabrication of periodic lattice-based cellular structures*. Computer-Aided Design and Applications. 13. 10.1080/16864360.2015.1059194.

[14] K. Raz, Z. Chval, F. Sedlacek, 2019. *Compressive Strength Prediction of Quad-Diametral Lattice Structures*. 69-74 pp., Key Engineering Materials

[15] Wikipedia contributors, "*Young's modulus*" Wikipedia, The Free Encyclopedia, https://en.wikipedia.org/w/index.php?title=Young%27s_modulus&oldid=940567807 (Acessed February 26, 2020).

[16] Egan, Paul & Gonella, Veronica & Engensperger, Max & Ferguson, Stephen & Shea, Kristina. (2017). *Computationally designed lattices with tuned properties for tissue engineering using 3D printing*. PLOS ONE. 12. e0182902. 10.1371/journal.pone.0182902.

AUTHOR INDEX